# Data Science, Data Visualization, and Digital Twins

*Edited by Sara Shirowzhan*

Published in London, United Kingdom

IntechOpen

*Supporting open minds since 2005*

Data Science, Data Visualization, and Digital Twins
http://dx.doi.org/10.5772/intechopen.87794
Edited by Sara Shirowzhan

Contributors
Johanna Schmidt, Yunus Egi, Engin Eyceyurt, Vung Pham, Tommy Dang, Nuno Alpalhão, Miguel de Castro Neto, Marcel Motta, Suzanna Long, Jacob Hale, Steven Corns, Vinayaka Gude, Paweł Kamiński, Piotr Kalinowski, Oskar Długosz



First published in London, United Kingdom, 2022 by IntechOpen
IntechOpen is the global imprint of INTECHOPEN LIMITED, registered in England and Wales, registration number: 11086078, 5 Princes Gate Court, London, SW7 2QJ, United Kingdom
Printed in Croatia

British Library Cataloguing-in-Publication Data
A catalogue record for this book is available from the British Library

Additional hard and PDF copies can be obtained from orders@intechopen.com

Data Science, Data Visualization, and Digital Twins
Edited by Sara Shirowzhan
p. cm.
Print ISBN 978-1-83962-943-3
Online ISBN 978-1-83962-944-0
eBook (PDF) ISBN 978-1-83962-945-7

For EU product safety concerns:
IN TECH d.o.o., Prolaz Marije Krucifikse Kozulić 3, 51000 Rijeka, Croatia, info@intechopen.com or visit our website at intechopen.com.

# Meet the editor

Dr. Sara Shirowzhan is a lecturer at the School of Built Environment (BE), University of New South Wales (UNSW), Sydney, Australia, where she teaches the City Analytics and Construction programs. She also serves as the co-chair of BE's Smart Cities and Infrastructure Cluster. Dr. Shirowzhan works as tomorrow's leading champion for the Chartered Institute of Building (CIOB). Her research interests include sensing technologies, enhanced GIS, BIM, digital twins, and artificial intelligence in technologies pertinent to BE informatics. She teaches and supervises students at UNSW in the areas of GIS, BIM, digital twins, AI, machine learning, city analytics, urban informatics, smart cities, infrastructure, construction informatics, and other relevant topics. She now serves on the editorial boards of the journals *MDPI* and *Advances in Civil Engineering*. She is also a topic board member of the *ISPRS International Journal of Geo-Information* as well as *Buildings*. Dr. Shirowzhan received her Ph.D. in Geomatics Engineering from the School of Civil and Environmental Engineering, UNSW.

# Contents

# Preface

Data science, data visualisation, and digital twins are trending in many disciplines. Appropriate data visualisation and analytics, enabled by data science, are required for informed decision-making in a variety of sectors. If expertise in advanced data analytics techniques are available, advanced data analytics approaches such as Artificial Intelligence (AI) and real-time, web-based, and interactive visualisations are used.

Advanced data visualisation methods, such as 3D, 4D, and so on, as well as dashboards, are great tools for better communication with stakeholders, better understanding and modelling of the current situation, forecasting future trends, and digital twinning of buildings, urban neighbourhoods, infrastructure, and cities in smart cities and built environments.

Professionals, academics, managers, planners, and policymakers have discovered that improved analytical methods of data science, such as machine/deep learning or AI, promise to provide superior insights from data, allowing them to make more educated decisions. In organisations, web-based systems that visualise such insights allow rich interactions among team members, clients, project managers, and stakeholders.

This book highlights established and advanced data science and visualisation technologies, given the benefits of data science, visualisation, and digital twinning. This book is divided into three sections based on the overall themes of the chapters.

Section 1 addresses web and dashboard-based visualisations. In the first chapter of this section, Scanostics features are implemented in JavaScript to illustrate their usability in 2D, 3D, and higher dimensions on the Web. In this chapter, Pham and Dang begin by describing the mathematical definitions of these Scanostics features, then provide installation instructions and implementation scripts for using these features on multivariate data via their GitHub page. In the second chapter in this section, Alpalhao, Castro Neto, and Motta discuss the benefits of developing dashboards for better decision-making in smart cities and show off their developed dashboard for monitoring spatiotemporal mobility patterns and indicators during the COVID pandemic.

Section 2 deals with 3D modelling of trees using point cloud data and digital twinning in the mining industry. In their chapter, Egi and Eyceyurt offer a system that uses machine learning algorithms to reconstruct topography from point cloud data and utilizes 3D tree modelling in such an environment for mobile communications. They also highlight the usability of their sensor fusion technology in smart city applications. In the final chapter of this section, Kalinowski, Dlugosz, and Kaminski suggest a digital twin of a mining shaft and hoisting system utilising BIM models for project management process improvement.

Section 3 demonstrates better flood modelling as well as the predicted future growth of visual data science. In their chapter, Hale, Long, Gude, and Corns present a way for exploiting open data and deploying deep learning algorithms in a GIS framework to anticipate flood inundation profiles for planners to employ. In the final chapter, Schmidt outlines the successful application of data visualisation algorithms in various data science workflows, compares data science libraries for various applications, and draws conclusions on the potential directions for future developments of advanced data visualisations.

**Dr. Sara Shirowzhan**
University of New South Wales,
School of Built Environment,
Sydney, Australia

Section 1

# Web/Dashboard Data Visualisation

Chapter 1

# JavaScript Implementation of Scagnostics and Its Applications

*Vung Pham and Tommy Dang*

## Abstract

Scagnostics is a set of features that characterizes the 2D distributions in the underlying data. Various real-world applications have been using Scagnostics visual features to detect unusual bivariate data correlations. Concomitantly, many applications are required to be implemented on web platforms due to their accessibility and convenience. Therefore, this chapter discusses a recent JavaScript implementation of Scagnostics, an extension to higher dimensional data, and its applications in detecting abnormalities in bivariate and multivariate time series data. Its implementation in JavaScript supports the tremendous demand for visual features in the web environment. Likewise, its higher dimensional implementations allow generating Scagnostics features for the rapidly growing multivariate data. Finally, conventional ScagnosticsJS computations involve time-consuming algorithms, and they are sensitive to slight changes in the underlying data. Therefore, this chapter also discusses a recent attempt to tackle these issues using machine learning to estimate the Scagnostics scores.

**Keywords:** Scagnostics, 3D Scagnostics, nD Scagnostics, JavaScript, Visual Features for Scatterplots, Data Correlation, High dimensional Data Analysis

## 1. Introduction

Visualizations on the web platform are getting popular because they are likely to be viewed on various devices without making any complicated software setup. A critical aspect of data visualization is detecting and highlighting the visual features in the underlying data. Regarding visual features, Scagnostics [1] is a popular set of features that allows characterizing the distribution of the underlying data. Though its two-dimensional (2D) version is prevalent, three-dimensional (3D) and higher-dimensional (nD) Scagnostics uses are limited.

There are several reasons for this limitation. First the 2D version of Scagnostics implementations were limited in programming languages like R [2], Python [3], and Java [4]. Furthermore, there was only one attempt at implementing Scagnostics in 3D space [5], and it is also supported in Java only. These programming languages encumber the uses of these visual features for the web, which is gaining popularity due to its convenience of use. Moreover, the lack of nD version nukes their uses for multivariate data in many application domains. Examples of such areas are those that often require multivariate data for better reliability and statistically sound analysis.

This chapter discusses a recent work called ScagnosticsJS [6]. It is a Scagnostics library implemented in JavaScript to support visualizing data features on the web.

Furthermore, it is the first officially published library that extends the Scagnostics features into 3D and nD. This chapter also discusses the uses of 2D and nD versions of ScagnosticsJS in detecting abnormalities in bivariate as well as multivariate time series data, namely Outliagnostics [7] and MTSAD [8], respectively. Finally, conventional Scagnostics computations involve algorithms that are time-consuming and sensitive to slight changes in the underlying data. Thus, this chapter's last section outlines an attempt to tackle these issues using machine learning, called ScagCNN [9]. Its main idea is to train and use the learned machine learning model to predict the Scagnostics scores instead of executing the conventional algorithms.

## 2. 2D Scagnostics

In 2005, Wilkinson [1] proposed and implemented a set of the graph-theoretic summaries of two-dimensional scattered point data, called Scagnostics. Since then, Scagnostics has been used in various application domains, such as high-dimensional time series analysis [8], clustering of scatter plots [4], and abnormality detection [7], to name but a few. This section summarizes the nine Scagnostics measures and their visual characteristics. We also refer interested readers to the original paper [1] for further details. These scores are broadly categorized into density, shape, and association measures.

There are five Scagnostics measures that are computed using density geometric features, namely: *outlying*, *skewed*, *sparse*, *clumpy*, and *striated*. They are characterized by the distribution of the edge lengths of the minimum spanning tree (MST) built on the 2D scatter points. Next, there are three Scagnostics measures that represent the shape geometric feature of scattered point data, namely *convex*, *skinny*, and *stringy*. Besides the MST, the shape measures also leverage the alpha ($A$) and convex ($H$) hulls built on top of the scattered points. Lastly, the association measure represents the symmetric measure of association between the two variables involved. The only one Scagnostics score in this category is *monotonic* score. Scagnostics uses the squared Spearman correlation coefficient of the 2D data to compute this score. We refer interested readers to the original work [10] regarding the formal definitions and how to compute these 2D Scagnostics scores.

There are several Scagnostics implementations in different programming languages such as R [2], Python [3], and Java [4]. These implementations are for 2D Scagnostics, and there is only one attempt to implement Scagnostics in application to 3D data [5]. ScagnosticsJS [6] is the first officially published Scagnostics implementation in JavaScript. It also extends the 2D Scagnostics measures for 3D and higher dimensional (nD) data points. The remained sections of this chapter discuss ScagnosticsJS implementations, its extensions, and its applications.

## 3. ScagnosticsJS: 2D, 3D, and nD Scagnostics for the web

This section describes the 2D, 3D, and nD Scagnostics implementation called ScagnosticsJS [6]. It is a library for Scagnostics in JavaScript, which is gaining popularity for *visualization for the web* [11]. Besides 2D Scagnostics, it is also the first officially published extensions of Scagnostics to 3D and nD. **Figure 1** shows the nine Scagnostics measures and their exemplar plots in 2D (a), 3D (b), and nD (c). The heatmaps next to these plots show the corresponding Scagnostics scores computed using ScagnosticsJS. Notably, ScagnosticsJS gives high scores (the highlighted, black bordered cells at the diagonals of the heatmaps) for the Scagnostics types that the scatterplots were generated to flag.

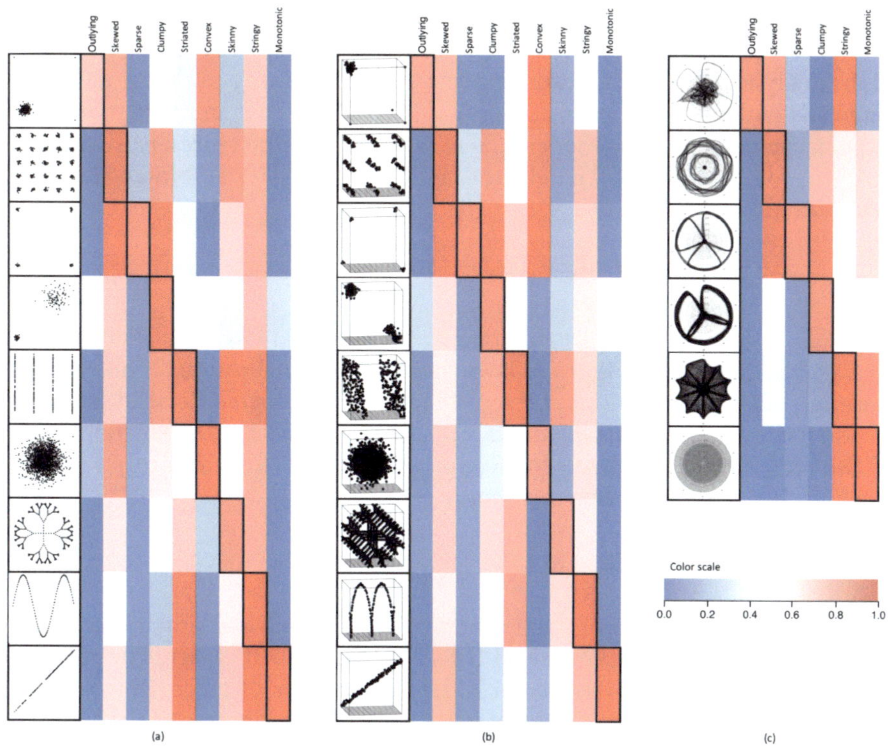


**Figure 1.**
*Scagnostics measures and their exemplar plots in 2D (a), 3D (b), and nD (c). The heatmaps show corresponding Scagnostics scores computed using ScagnosticsJS. Highlighted, black-bordered cells at the diagonals indicate that ScagnosticsJS flags high Scagnostics scores for their corresponding typical Scagnostics plots.*

## 3.1 Binning

The Scagnostics computation starts with the binning process. This step reduces the computation expense when handling a large number of data points and allows more stable Scagnostics computations. Rectangular binning is a simple and popular binning method in data science in general. However, hexagon and leader binning algorithms are two commonly used methods for Scagnostics computation. Though both algorithms are similar in terms of time complexity, leader binning has its advantage of preserving the underlying data's original shape. Contrariwise, hexagon binning does not work well with Box Plot Rule [7] because the distance between neighboring hexagons is always the same due to the binning shape and their consecutive arrangement.

**Figure 2** depicts the difference between leader binning (d) and its rectangular (b) and hexagon (c) counterparts on the same original data (a). Each leader's size

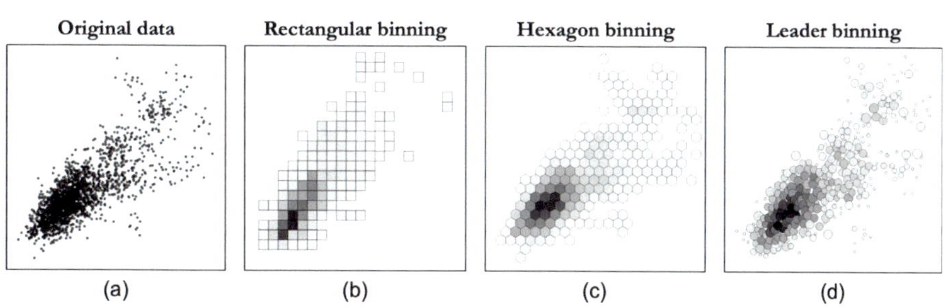


**Figure 2.**
*Original data (a) and binning methods: Rectangular binning (b), hexagon binning (c), and leader binning (d).*

indicates its coverage, while its intensity highlights the number of points that fall into that bin. Each binning algorithm has its pros and cons depending on the data and the analysis task, so ScagnosticsJS provides the flexibility to choose either hexagon or leader binning in its 2D version. It is hard to find an appropriate shape for the hexagon in nD, but a hyper-sphere is an appropriate representation in nD equivalent to a circle in 2D. Therefore, ScagnosticsJS uses leader binning for its nD implementation.

### 3.2 ScagnosticsJS 2D implementation

The 2D ScagnosticsJS implementation includes several intermediate computation stages as depicted in **Figure 3**. They include 1) normalization, 2) binning, 3) generating triangulation, 4) computing MST, 5) finding degree 1 and 2 vertices, and 6) finding convex and concave hulls. Since 2D Scagnostics implementation is popular, we refer interested readers to these original papers [1, 6] for further details regarding these stages' implementations.

### 3.3 ScagnosticsJS 3D implementation

Our 3D implementation of Scagnostics in ScagnosticsJS bases mainly on how corresponding geometries transferred from 2D space into 3D space. The concept of the MST is exactly the same as that in 3D using Euclidean distance. Obviously, the *outlying*, *skewed*, *stringy*, and *clumpy* measures use the distribution of the lengths of the MST's edges. Thus, their 3D implementations can be naturally adapted from 2D. Additionally, *monotonic* score in 3D is the partial correlations of the three variables: $c_{monotonic} = \max\left[\rho^2_{X,Y|Z}, \rho^2_{X,Z|Y}, \rho^2_{Y,Z|X}\right]$.

*Striated* patterns in 2D involve parallel and smooth (e.g., spiral) lines [1]. The equivalent *striated* patterns in 3D involves all of these 2D patterns plus the parallel planes. So, the *striated* score is revised to account the angle between adjacent planes ($p$) formed by every three consecutive edges ($e_1$, $e$, and $e_2$) of the MST. Different from [5], we need to consider $|\cos\theta| > 0.75$ instead of $\cos\theta < -0.75$ as the 2D version. Therefore, the 3D *striated* score is now computed as:

$$c_{striated} = \frac{1}{|V|} \sum_{v \in V^{(\geq 2)}} I\left(\left|\cos\theta_{p(e,e_1)p(e,e_2)}\right| > 0.75\right) \tag{1}$$

where $V^{(\geq 2)} \subseteq V$ are vertices of degree $\geq 2$. The $\cos\theta_{p(e,e_1)p(e,e_2)}$ is calculated as the dot product of the unit normal vectors of the two planes $p(e, e_1)$ and $p(e, e_2)$. These unit normal vectors are, in turn, computed using the cross products of the vectors made of source and target nodes of $e$ and $e_1$ for the first plane and $e$ and $e_2$ for the second plane.

**Figure 4** shows the synthesized samples generated to justify the use of $|\cos\theta| > 0.75$ instead of $\cos\theta < -0.75$. Specifically, these samples include two

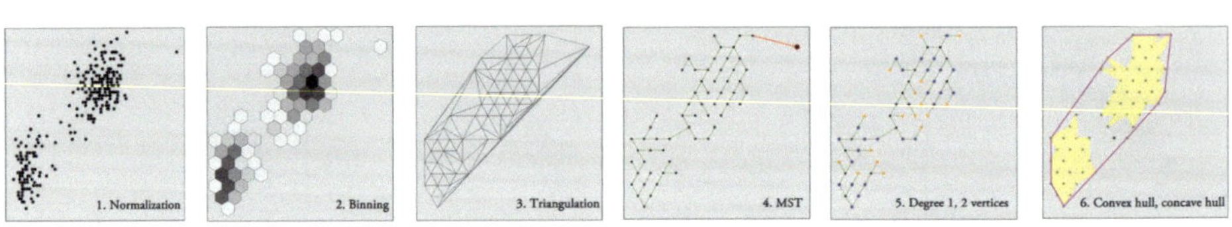


**Figure 3.**
*Main algorithms used in 2D Scagnostics computation: 1) normalization, 2) binning, 3) triangulation, 4) MST (with a red outlier), 5) vertices with degree 1 (orange) and 2 (blue), and 6) convex and concave hulls.*

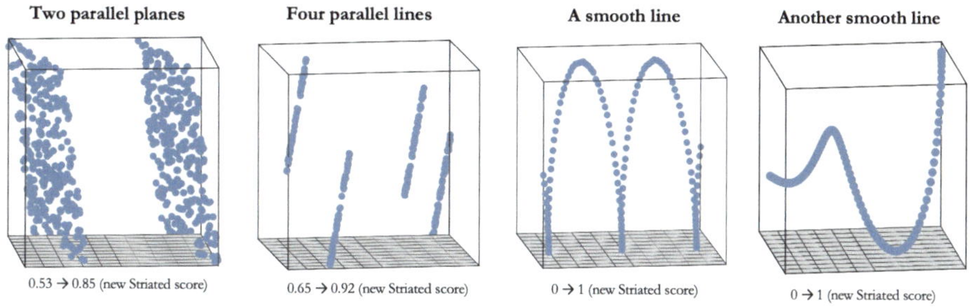


**Figure 4.**
*Comparing striated scores using* $\cos\theta < -0.75$ *vs.* $|\cos\theta| > 0.75$ *thresholds for two parallel planes, four parallel lines, a smooth line, and another smooth line to test the generality of the proposed formula.*

parallel planes, four parallel lines, and two smooth lines (the last two cases). The *striated* scores for the first two cases are relatively similar and are reasonable. However, in the third case, using $\cos\theta < -0.75$ threshold leads to *striated* score of 0 versus 1 when using $|\cos\theta| > 0.75$ threshold. As of the *striated* pattern definition, this score of a smooth line should be 1. We also tested the proposed change to other types of smooth lines to ensure that this formula is generalized to measure the smooth lines' striated visual feature. The last case is another example of such many smooth lines that we had tested on.

The 2D *c onvex* and *skinny* scores leverage the perimeters and areas of the convex and concave hulls built on top of the underlying data. In 3D, the 2D lines are equivalent to planes, and the $\alpha$ radius for the circle used to calculate the 2D alpha hull is that for the sphere in 3D used to calculate the 3D alpha hull. Therefore, 3D version convex and skinny scores use the equivalent surface areas and volumes of these shapes in 3D, instead of perimeters and areas as in 2D. These scores are formally defined as:

$$c_{convex} = volume(A)/volume(H) \tag{2}$$

$$c_{skinny} = 1 - \sqrt[6]{36\pi}\sqrt[3]{volume(A)}/\sqrt{surfacearea(A)} \tag{3}$$

where $H$ and $A$ are the convex and alpha hulls in 3D space, respectively. Also, the constant $\sqrt[6]{36\pi}$ is to assure that a sphere has $c_{skinny} = 0$.

The area of a 3D hull is the sum of areas of all triangles of that hull. These triangles are the faces of the hull as the result of the hull computation. Also, ScagnosticsJS uses the algorithms from Robert Nürnberg [12] to efficiently compute the volume of a 3D hull. Algorithm 1 summarizes the steps for computing the volume of a 3D hull from its faces. It is worth noting that this algorithm works for both convex and concave hulls.

**Algorithm 1** Compute the volume of 3D hull from its set of faces.

1: **procedure** COMPUTEVOLUME(*faces*).
2: initialization: $n$ = number of faces, $volume = 0, i = 0$.
3: **while** $i < n$ **do**:
4: $triangle_i = faces[i]$
5: $vector_i$ = one vertex of $triangle_i$.
6: $\hat{n}$ = normal vector of $triangle_i$.
7: $volume = volume +$ dot product of $vector_i$ and $\hat{n}$
8: $i = i + 1$
9: **return** *volume*.

Lastly, we defer the discussion regarding the computation of *sparse* score to the next section. The reason is that this measurement deserves further discussions and is generalized from 2D to 3D and naturally to nD versions.

### 3.4 ScagnosticsJS nD implementation

Like the 3D implementation, the nD version depends on how geometries in 2D and 3D spaces are translated into nD. Specifically, regarding building the MST in nD space, besides Euclidean distance, one should also consider using other metrics such as Manhattan distance metric ($L_1$ norm) or $L_k$ norm where $k$ is a fraction. Asides from the distance metric consideration, the MST computation remains the same, so do the computations of the *outlying*, *skewed*, *sparse*, *clumpy*, and *stringy* scores. Also, ScagnosticsJS uses maximum partial monotonicity among all pairs of variables as the *monotonic* score in nD.

With values normalized to the range $[0, 1]$ as described in the 2D implementation section, in nD space, the maximum distance between any two points could get up to $n^{1/k}$, where $n$ is the number of dimensions, and $k$ is the value for the distance metric $L_k$ used. For instance, in case of Euclidean distance ($k = 2$), this distance is $\sqrt{n}$. This case is an extreme one because, at most, there is only one edge length with such value. However, the fact is that the higher the number of dimensions, the higher the possibility that the MST edges get longer than 1. Therefore, using $q_{90}$ of the MST edge length distributions as the *sparse* score does not generalize to 3D or nD versions.

**Figure 5** shows 6D, 10D, and 16D sets of data points synthetically generated to illustrate this circumstance. If we use the $q_{90}$ as a sparse measure, the three plots have sparse scores as 1.76, 2.16, and 2.80, respectively. There are two issues with these scores. First, they are out of the Scagnostics score range (0 to 1). Second, the three synthesized plots are visually similar, and they only differ in numbers of dimensions. In other words, the higher sparse scores of the higher dimensional plots are due to the higher number of dimensions they have, not due to their intrinsic visual features. Therefore, ScagnosticsJS proposes the sparse score as:

$$c_{sparse} = q_{90} / \sqrt{\left\lfloor \frac{2 \times n}{3} \right\rfloor} \quad (4)$$

where $n$ is the number of dimensions, and we use the floor operation ($\lfloor\rfloor$) because MST always selects the lower distances first. The numerator 2 is due to the pairwise distance between points. The denominator 3 is the requirement that we would like to have at least 3 MST edges with lengths greater than or equal to the $q_{90}$. Also, this is compatible with the cases $n = 2$ in 2D implementation. This correction

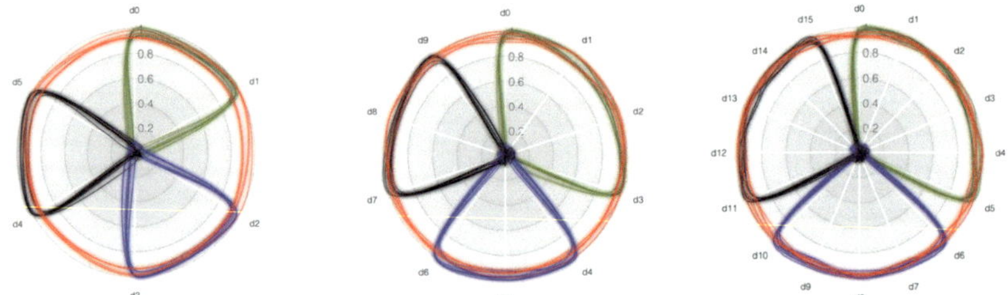

**Figure 5.**
*Synthesized 6D, 10D, and 16D plots with high sparse scores: 1.76, 2.16, and 2.80 if we use the $q_{90}$ measure, though they are visually similar. Radar chart, in this case, is one of many ways to represent multivariate data, and line colors represent different classes.*

factor normalizes the sparse scores for the scatter plots in **Figure 5** into approximately 0.88 (0.880, 0.882, and 0.885, respectively) in these three cases.

Currently, ScagnosticsJS does not implement the nD *convex* and *skinny* scores due to their limited utility [13]. Furthermore, these scores depend on the convex/alpha hull in nD space, and the computations of these shapes pose performance constraints. Specifically, even with an approximation approach such as one from [14], the complexity of hull calculation is still $\mathcal{O}\left(N^2 m^{3/2} \log \frac{m}{\epsilon_0}\right)$, where $m$ is close to the number of vertices of approximation and $\epsilon_0$ is the maximum error. The time complexity makes it impractical to incorporate these scores in the current nD version.

Similarly, current ScagnosticsJS version does not provide the implementation for *striated* score in nD space. There are several reasons for this. First, calculating this score involves computing the angle between every two consecutive MST edges or two planes formed by every three consecutive MST edges in 2D and 3D cases, respectively. This calculation further involves the cross-product of two vectors. This cross-product does not exist in the space with dimensions higher than three. In other words, there are infinitely many unit vectors orthogonal to any given two. The second reason is also about utility. Specifically, lines and planes can be found in higher-dimensional space, but there is not often much reason to use them [15].

Interested readers can refer to the Github page of ScagnosticsJS, at https://idatavisualizationlab.github.io/ScagnosticsJS, to explore and learn how to use this library in readers' application domains. The next section describes two typical recent applications of this library to demonstrate the uses of 2D and nD Scagnostics to extract visual features for visualizations on the web platform.

## 4. Applications of ScagnosticsJS

JavaScript implementation of Scagnostics facilitates the use of these measures on the web. This section describes two recent works called Outliagnostics [7] and MTSAD [8] as the typical applications of 2D and nD ScagnosticsJS, respectively.

### 4.1 Visualizing temporal discrepancies in outlying signatures of data entries

Outliagnostics [7] is a recent, typical application of 2D ScagnosticsJS. It is an application of this library to analyzing 2D temporal data concentrating on detecting data entries that are significant in contributing to the outlying score of a scatterplot. Its prototype also supports interactive explorations of abnormalities in large time series. It uses ScagnosticsJS to calculate how much a data point adds to the outlying score of the underlying scatter points as a whole at every time point. In some cases, outliers can be detectable using one-dimensional data. However, in many other cases, they are only detectable in multidimensional space. Therefore, instead of using conventional approaches such as Box Plot Rule to find outliers in one-dimensional data, ScagnosticsJS allows determining the outliers using two-dimensional data points.

**Figure 6** demonstrates the use cases of 2D outlying detection where these outliers in any marginal projections are indiscernible. **Figure 6(a)** is the scatterplot of international debt data (*debt* vs. *population*) in 2017: Each data point is a country in the example scatterplot. One can rely on the *debt* axis alone to determine that Pakistan and China are outliers, or the *population* axis alone to discern China and India as outliers. There are similar cases in **Figure 6(b)** for the New York stock exchange in January 2011 (*price* vs. *volume*), such as BAC, AAPL, GOOG. However, it is relatively hard to tell if NFLX is outlying using either one of the axes in this scatterplot. Similarly, it is even harder to tell if Iraq, El Salvador, or Iran in **Figure 6(c)** using either *Female* axis or *Male* axis alone in this *Life expectancy in 1982* scatterplot. However, they are visually outliers if both axes are taken into account.

While analyzing the outlying impact of data points to the overall scatterplot outlying score, besides the outliers, Outliagnostics also considers "inlying" data points. Specifically, it defines an "inlier" as an observation that lies in the interior of statistical distribution, and its absence makes the detection of other outliers easier or possible. For instance, *A* and *B* are two data entries in a distribution. *A* is an inlier if removing *A* makes *B* an outlier. Outliagnostics's approach to identifying the outlying contribution of an individual data point to its overall dataset is to compute the difference of the outlying scores while having and not having that particular data point in the underlying dataset. This method is called the *leave-one-out* approach.

This *leave-one-out* approach allows estimating the outlying contributions of individual data points to the overall scatterplot as a whole. This approach is computationally intensive. However, Outliagnostics leverages binning and parallel computing to achieve a near-linear computation complexity concerning the size of a dataset. Specifically, the leave-one-out approach is "selective" because it only leaves out singleton bins. The reason is removing an observation from a dense bin will not impact the Scagnostics outlying scores.

**Figure 7** shows the main components of Outliagnostics. On the left (a) is a control panel that shows color legends and allows users to set various display properties such as ordering fields or searching for items, in case needed. The top bar (b) represents the time series going from left to right. Right below that time series bar is the set of thumbnails (c) for the 2D data represented as scatterplots over time. These scatterplots' backgrounds represent their outlying scores (red background means higher outlying score while the blue one indicates low outlying score).

The thumbnails below the tag clouds (d) visualize the five data entries with the highest contributions on the overall outlying score (increasing or reducing). The text colors in these tag clouds designate the inlying (green) or outlying (purple) contributions of the corresponding entries. Also, Outliagnostics provides a customized box-plots view (e), which gives an overview of the outlying and inlying information of the scatterplots at individual time steps. A stream graph overlayed on top of these box-plots shows how these outlying/inlying values evolve.

Finally, Outliagnostics also provides a visualization section called instance profiles, as shown in **Figure 7(f)**. This section allows users to investigate how outlying/inlying values evolve at the individual instance level. Specifically, each item has a base-line representing the outlying score of the overall scatterplot. Suppose at a time

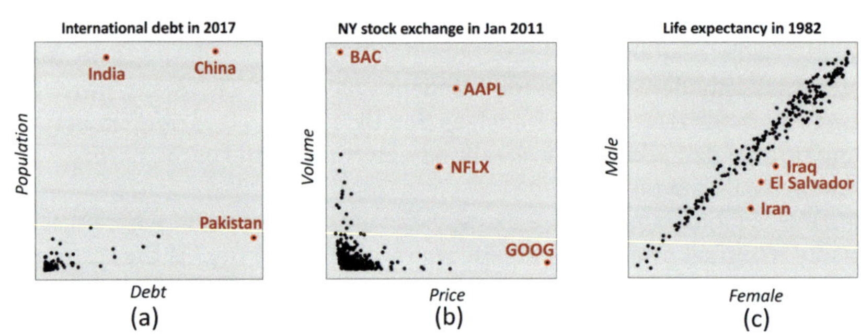


**Figure 6.**
*Examples of two-dimensional outliers that might not be detectable in individual dimensions: (a) international debt, (b) New York stock exchange, and (c) life expectancy data from the World Bank [7].*

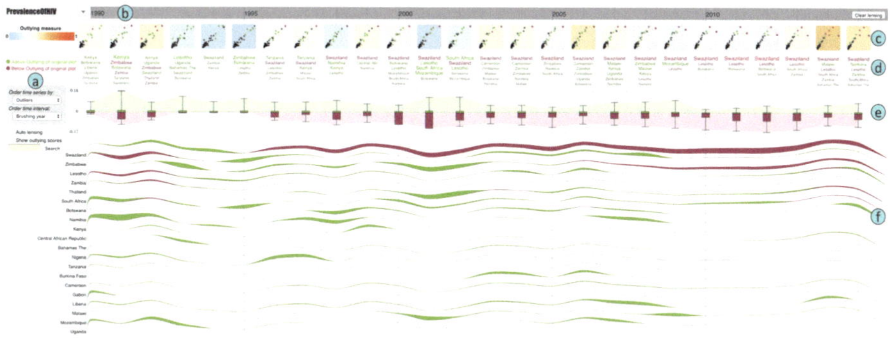


**Figure 7.**
*Outliagnostics visualization components: (a) control panel, (b) lensing area, (c) scatterplot series, (d) top countries clouds, (e) customized outlying boxplots, and (f) outlying profiles [7].*

point the instance contributes more to the inlying/outlying score. In that case, there are green streams above this base-line or purple streams below it to represent the changes of the outlying scores when leaving the current data item out, correspondingly. The streams' heights denote the inlying/outlying difference when this specific instance is present or absent from the scatterplot. The use of stream graphs for these scores allows the users to observe how the outlying scores evolve for a particular instance.

**Figure 8** depicts a use-case of Outliagnostics applies to the World Life Expectancy dataset retrieved from UIC repository [16]. This dataset is the life expectancy (male vs. female) from 263 countries worldwide in 56 years. The time-series boxplots and the outlying/inlying streams highlight the 1970s, early 1990s, and late 1990s - early 2000s as three periods with high outlying scores. The time series shows Cambodia, Rwanda, and Sierra Leone as the three profiles that are ranked on top of the list with thicker outlying streams. In other words, they contribute more to the overall outlying scores over these time periods, correspondingly. Clicking on the thumbnails scatter plots at each of the periods shows their detailed views in boxes (a), (b), and (c), respectively. Moreover, users can mouse-over these three countries in these detailed views to inspect the male and female life expectancy for these countries. Specifically, Combodia had as low as 27 (male) and 21 (female) in 1975. These numbers are 30 and 26 for Rwanda in 1992. Similarly, They were 38 and 36 for Sierra Leone in 1998. These low values were caused by the 1978–1991

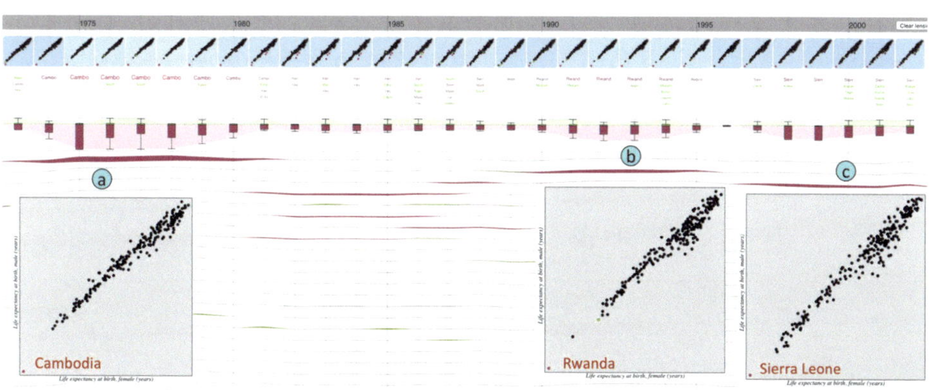


**Figure 8.**
*Outliagnostics highlights Cambodia, Rwanda, and Sierra Leone as outliers in the 1970s (a), early 1990s (b), and late 1990s - early 2000s (c), respectively.*

Cambodian Civil War, the 1990–1994 Rwanda Civil War, and the 1991–2002 Sierra Leone Civil War, respectively.

Finally, source codes, video demo, and web prototypes of Outliagnostics are available from this project's Github page at https://outliagnostics.github.io/index.html.

### 4.2 Multivariate time series abnormality detection and visualization

Several outlying items are not detectable using a lower number of dimensions, but they are in higher dimensional space, as discussed in Section 4.1. Therefore, this section describes another recent application of the nD version of ScagnosticsJS [6] that applies to detecting abnormalities within a multivariate time series using a web-based application, called MTSAD [8]. This application is a natural extension of Outliagnostics to the nD data. Specifically, Outliagnostics claims that several outliers are not detectable using an individual variable, but they are evident when both variables of the bivariate data are taken into account. Similarly, one may see the same argument for 3D and nD space.

**Figure 9** shows the main interface of MTSAD that monitors nine critical variables in a high-performance computing center [17]. Similar to Outliagnostics, MTSAD also contains a time-line, plot preview thumbnails over time, the customized box-plots depicting how overall outlying/inlying scores evolve, the item profile sections, and a control panel with interactive options to support exploration during the abnormality detection process. The same leave-one-out strategy, as described in Section 4.1, is used to calculate the outlying contribution of an individual data instance to the overall outlying score of the whole set (i.e., of all involving data items) at a specific time step.

There are two main differences between Outliagnostics and MTSAD. First, the former uses 2D ScagnosticsJS while the latter utilizes the nD version to calculate the outlying scores for individual data-plot at each time step. The second difference is that MTSAD leverages small-multiples (radar-charts) to show the multivariate time series instead of showing scatterplot for 2D data points as in Outliagnostics. Each small-multiple visualizes the monitoring variables at a time step. Each radar-chart might potentially need to render a large number of data entries (467 computation nodes in this case). This large number of paths indicates a high rendering time and produces visually cluttering issues. Concomitantly, a higher level of data and visualization abstractions (such as grouping and visualizing a group of data instead of individual data points) gives a better overview of the data and faster rendering time [18]. Therefore, this application only displays individual inlying (green paths) and outlying (red paths) entries, detected by the nD version of ScagnosticsJS with the leave-one-out approach. It then uses k-means algorithm [19] to aggregate other

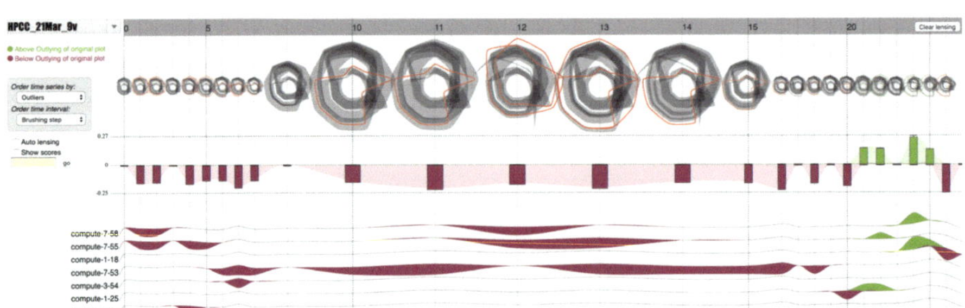


**Figure 9.**
*MTSAD primary visualization components for a multivariate time series dataset from a high performance computing center: 1) a time-line, 2) radar-chart small-multiples, 3) overall outlying box-plots, 4) the data-entry profile, and 5) a control panel to assist the abnormality exploration interactively [8].*

non-inlying/non-outlying computing nodes into clusters and visualizes them as gray bands to reduce rendering time. This approach highlights outlying/inlying observations in the underlying data, which assists the detection of abnormality.

**Figure 10** shows a snapshot of MTSAD applies to monitoring nine essential variables from a high-performance computing center. The monitoring period is from 2:00 PM until 4:25 PM on March 21, 2019 (in 30 time-steps with five minutes each). **Figure 10(a)** shows a detailed view of these monitoring CPU-health-related metrics at 2:20 PM. Notably, some nodes were heated, and two of them were outlying because they have high values for temperature metrics. Also, some other nodes detected the heat increment and increased their fan speeds. Since then, these two metrics raised rapidly.

As depicted in **Figure 10(b)**, By 3:50 PM, our visualization highlighted that the recorded values for these two variables were high for many computation nodes. Finally, our monitoring system could not receive data from the computation nodes at around 4:25 PM. MTSAD detected and reported this suspicious behavior to the high-performance computing center's administrators. They acknowledged the matter and clarified that the cooling system's fault operations (chilled water system) caused the issue. The monitoring system did not receive any signal by 4:25 PM because the system administrators performed an emergency shutdown to circumvent any further harm to the computation facility.

Furthermore, this use-case also suggests that outliers and inliers are equally important in our abnormality detection approach. As shown in **Figure 10(b)**, the Box-Plot rule cannot detect outlying data entries at the 3:20 PM time step. Notably, at this time, *compute-1-25* and *compute-7-55* were at their extreme values (one had very low and another very high values for the tracking CPU-health metrics). The masking effect hides the outlying characteristics of these two instances. Specifically, leaving one of the two nodes out, the remained data points will have a high Scagnostics outlying score. That said, inliers allow us to detect potential outliers. Consequently, identifying the inlying measurements permits us to tackle the masking effect.

Finally, the source codes and web prototype of MTSAD application are available at https://idatavisualizationlab.github.io/V/MultiOutliers.

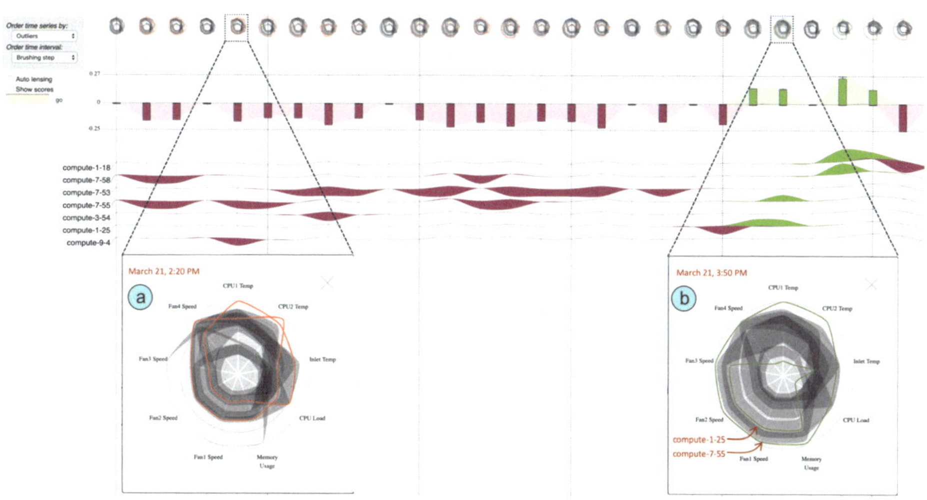


**Figure 10.**
*MTSAD visualizations applied to monitoring essential variables at a high performance computing center. From 2:00 PM until 4:25 PM on march 21, 2019. There was a chill water issue during this period. At 2:20 PM (a), the computation nodes' temperatures increased, and a few nodes sensed the heats and pumped their fan speeds (panel a). At 3:50 PM (b), other computation nodes also increased their fan speeds after sensing the heats [8].*

## 5. Estimating visual characteristics of 2D scatterplots via CNN

Computing individual Scagnostics scores for individual scatterplot is reasonably fast (in terms of milliseconds) [6]. However, it is still relatively slow for applications that need to do the Scagnostics computations extensively such as Outliagnostics [7] and MTSAD [8] for their real-time monitoring purposes. For these scenarios, the Scagnostics computation is still considered as having computationally expensive algorithms. Moreover, these algorithms are sensitive to the slight changes in the underlying data distribution within the scatterplot [9]. Concomitantly, with its recent advancements, Convolutional Neural Network (CNN) is gaining traction and has several state-of-the-art results in computer vision tasks. Therefore, ScagCNN [9] proposes to use CNN models to estimate the Scagnostics scores (prediction model) and classify set of scatterplots to typical Scagnostics types (classification model). ScagCNN aims to improve the Scagnostics computation time and reduce the sensitivity to the small shifts in the distribution of the underlying data. This section describes the architecture of ScagCNN, the configurations of its prediction and classification CNN models, and their accuracy.

### 5.1 ScagCNN architecture

**Figure 11** depicts ScagCNN's schematic overview. Specifically, it uses real-world and synthesized data to train, validate, and test the CNN prediction and classification models. Using real-life datasets allows our solution to work in practice. In comparison, the synthesized one provides the ScagCNN with ground-truth labels while training classification models to detect typical Scagnostics types (the nine measures). In other words, these artificially generated plots have labels required to train the ScagCNN classification model. Also, ScagCNN uses ScagnosticsJS library [6] to compute the Scagnostics scores for these training, validating, and testing datasets.

The scatterplot data is converted into binary images and used as inputs for ScagCNN. Specifically, it also normalizes every variable's values into a unit range (i.e., [0, 1]). The normalized data then undergoes the binning step with 40 × 40 binning resolution (as suggested by the original Scagnostics paper). There are several reasons for doing this. First, normalizing and binning are the two data preprocessing steps done in the conventional approach. Please refer to Section 3 for why these steps are essential. Moreover, CNN models often require their inputs to have the same shape. Therefore, ScagCNN sets the binning resolution to 40 × 40, as mentioned in the original Scagnostics's paper. Notably, Scagnostics's computation algorithms do not take the number of data points within an individual bin into account (it is a bin if there is at least a point that falls within it and not otherwise, careless of the number of points inside each bin). Therefore, it's natural to view the

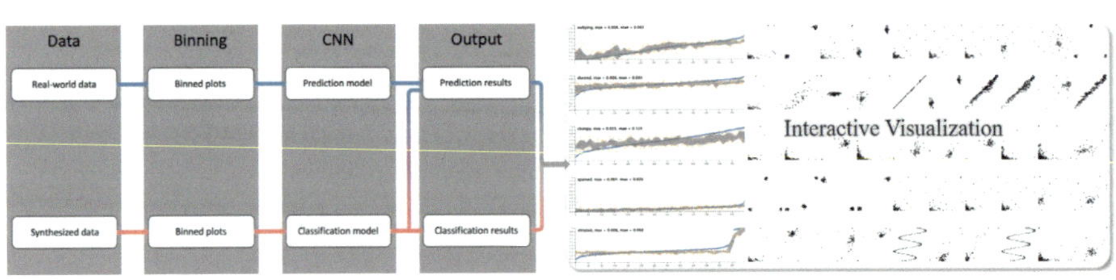


**Figure 11.**
*ScagCNN [8] main stages: 1) data processing 2) binning, 3) building CNN prediction and classification models, 4) using the learned models to give results.*

binned data as binary images when using them as inputs for the classification/prediction models.

ScagCNN models include a prediction model and a classification model. The former estimates the Scagnostics scores, while the latter classifies the underlying set of data points into a type that corresponds to one of the nine Scagnostics measurements. Both models use binary images, resulting from applying a binning method to the original data points as their inputs. The use of the synthesized dataset is necessary because it has the Scagnostics type as the ground-truth labels to train the classification model. The labels are the result of the generation process of this artificial dataset. Specifically, each generated set of data points is designed to have a high value for a specific Scagnostics measurement and is assigned to the corresponding Scagnostics type. Contrariwise, ScagCNN prediction model can leverage both real-world and synthesized data for its training, validation, and testing stages.

Artificial neural networks have high predictive power thanks to their freedom to learn abstract, salient features from the underlying data. This predictive power comes at the cost of explainability. In other words, the learned features are too abstract that humans cannot interpret their meanings. The lack of interpretability circumvents us from deploying our learned models into real-life since we cannot assure our models actually learn from appropriate features or merely memorize the underlying data's trivial characteristics. Consequently, there is a need to make them transparent [20]. ScagCNN accommodates the need for interpretability by offering users a web page. This web page visualizes the intermediate results of the algorithms involved in the Scagnostics computation. It also visualizes the features that the CNN models extracted in their various convolutional layers. These two tasks allow the users to debug and interpret the accuracy of the ScagCNN models. Specifically, the former will enable us to see if there is a bug in each of the underlying Scagnostics algorithms. Simultaneously, the latter allows the users to check if the extracted features are relevant or trivial. After having a thorough knowledge of these models' result generation process, users can make an educated decision regarding their accuracy.

### 5.2 ScagCNN prediction and classification models

Training an efficient CNN model involves experimenting with various architectures and tuning many hyperparameters. It is impossible to exhaustively search through many possible combinations of these configurations and hyperparameters to build a good CNN model. Consequently, a common approach in practice is to investigate and tune the existing CNN models appropriate for our specific domain (computer vision in this case). They are AlexNet [21], GoogleNet [22], and VGG-16 [23], to name but a few. These state-of-the-art models give excellent results in various computer vision tasks. However, in VGG-16, Visual Geometry Group (VGG) [24] provides an invaluable, reusable basic block for building CNN models following its standard [25], called a VGG block. Specifically, designers can stack one or more such blocks while building CNN models for their problem at hand. This block with guided architecture and hyperparameters helps reduce the number of possibilities that we need to search on while building our models.

Even using VGG as the building block, there are still many possibilities to experiment. Therefore, our approach starts with one VGG block and then stacks more VGG blocks and tests if more blocks result in improved validation accuracy. Suppose the training accuracy gets significantly higher than the validation one. In that case, we will start exploring techniques for overcoming overfitting, such as dropout, early-stopping, and weight regularization, and batch normalization. Once the overfitting issue is solved, we can continue stacking more VGG blocks and check if going deeper helps increase the performance. Finally, in this specific case, experiments show that dropout and early stopping help tackling overfitting issues while batch normalization and weight regularization do not. **Table 1** summarizes the experimented number of VGG blocks and overfitting techniques used while training ScagCNN models. Going from left to right means stacking more VGG blocks, and "*Y*/*N*" means the corresponding layer is used or not correspondingly.

Configuration 1 is the simplest one, which uses a single VGG block and results in a testing MSE of 0.0155. Configuration 2 adds one VGG block, which helps to reduce the MSE to 0.0136. Similarly, configuration 3 adds another VGG block. However, it does more harm than good due to overfitting. Therefore, after every VGG block and the fully connected layer (FC1), a dropout layer is used to tackle the overfitting issue. The training histories show that the overfitting is not devastating. Thus, we used 0.1 as the dropout rate. There are also attempts to then train with 1, 2, and 3 VGG blocks combined with dropouts in configurations 4, 5, and 6, respectively. Notably, having dropouts helps to improve the performance (validation MSEs reduce significantly). However, adding more than two VGG blocks reduces the model's performance. Lastly, training neural networks with batches of data may help reduce training time and improve generalization. ScagCNN achieves the former advantage via parallelization. The latter is due to the weight updates for batches use the average of their gradients generated on a group of data items instead of a single one (in stochastic case). Configurations 1 to 6 use a batch size of 64, and configurations 8, 9, and 10 are experiments with batch-size set to 'None.'

Finally, ScagCNN uses configuration number 8 in **Table 1**. This configuration has two VGG blocks (VGG1 and VGG2), two fully connected layers (FC1 and FC2), and three dropout layers. There is one dropout layer after every layer except for the output layer (FC2). Both VGG1 and VGG2 have a convolution layer with 32 and 64 filters of size ($3 \times 3$), respectively. Each of the VGG blocks also has a max-pooling layer with a pool size of ($2 \times 2$) to reduce the size of the feature map generated by every VGG block. FC1 and FC2 have 128 and nine hidden units, respectively. The 128 hidden units of FC1 is to increase the non-linearity, thus improves the model's predictive power. The nine hidden units in FC2 correspond to the nine Scagnostics measurements. Notably, though not depicted in the table, all the convolutional fully connected layers are followed by a ReLU [26] activation function. The classification model also follows configuration 8 in **Table 1**. Because this architecture can extract

| No. | VGG1 | DO1 | VGG2 | DO2 | VGG3 | DO3 | FC1 | DO4 | FC2 | Batch | MSE[a] |
|---|---|---|---|---|---|---|---|---|---|---|---|
| 1 | Y | N | N | N | N | N | Y | N | Y | 64 | 0.0156 |
| 2 | Y | N | Y | N | N | N | Y | N | Y | 64 | **0.0136** |
| 3 | Y | N | Y | N | Y | N | Y | N | Y | 64 | 0.0145 |
| 4 | Y | Y | N | N | N | N | Y | Y | Y | 64 | **0.0121** |
| 5 | Y | Y | Y | Y | N | N | Y | Y | Y | 64 | 0.0122 |
| 6 | Y | Y | Y | Y | Y | Y | Y | Y | Y | 64 | 0.0127 |
| 7 | Y | Y | N | N | N | N | Y | Y | Y | None | 0.0120 |
| 8[b] | Y | Y | Y | Y | N | N | Y | Y | Y | None | **0.0108** |
| 9 | Y | Y | Y | Y | Y | Y | Y | Y | Y | None | 0.0114 |

[a]*Mean Squared Error.*
[b]*Selected configuration.*

**Table 1.**
*Experimented CNN settings and their MSEs on the test dataset.*

salient features useful for the Scagnostics score predictions and then it should also be appropriate for the Scagnostics classification task. However, for the classification task, we replace the activation function of the output layer (FC2) to softmax [26] instead of ReLU.

### 5.3 ScagCNN prediction and classification results

**Table 2** summarizes the prediction results using the learned model to predict the Scagnostics scores on the testing dataset. "MSE" (mean-squared-error) and "MAE" (mean-absolute-error) are the two metrics that we use to measure the performance of the prediction model. Notably, the Scagnostics measurements have MAE less than or equal to 0.1, except clumpy. Scagnostics clumpy calculation depends on two edges in the runt statistics (one from the considering edge and another from the shorter sub-tree). Thus, this score is not robust to small changes in the underlying data [9].

Similarly, **Figure 12** depicts the sampled classification accuracy on synthesized and real datasets in panels (a) and (b), respectively. There are ground-truth labels for the synthesized dataset. Thus, it's straightforward to perform the accuracy evaluation on this set, and the accuracy is 98%. The high accuracy score may due to the generation bias of the synthetic data. Specifically, we generate the scatter plots of a type in such a way that it will flag high values for its corresponding Scagnostics measurement. Contrariwise, we do not have the ground-truth labels on the real dataset. Thus, we first sampled 100 scatterplots from the real-life test set, performed the classification on these plots, and asked two Scagnostics experts to evaluate the predictions' accuracy. The qualitative evaluation revealed that, out of 100 predicted labels, nine of them are incorrect. Notably, eight of them are related

| | Outlying | Skew | Clumpy | Sparse | Striated | Convex | Skinny | Stringy | Monotonic |
|---|---|---|---|---|---|---|---|---|---|
| MSE | 0.008 | 0.006 | 0.025 | 0.001 | 0.008 | 0.018 | 0.011 | 0.005 | 0.007 |
| MAE | 0.065 | 0.064 | 0.129 | 0.026 | 0.068 | 0.100 | 0.085 | 0.057 | 0.063 |

**Table 2.**
*Mean Squared Error (MSE) and Mean Absolute Error (MAE) for 9 Scagnostics scores using conventional Scagnostics algorithms vs. ScagCNN predictions.*

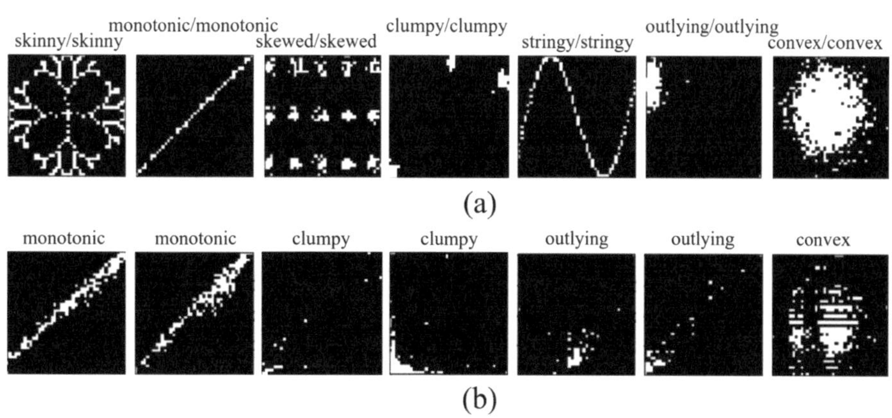


**Figure 12.**
*Classification results for randomly selected scatterplots from synthesized and real datasets. Panel (a) depicts the classification results on the synthesized test set. The title is in the form of* ground-truth/predicted *labels. Panel (b) shows the predicted labels for the real-life test data [9].*

| Task | Time (ms) |
|---|---|
| Average scagnostics time per plot | 20.019 |
| Scagnostics time per plot standard deviation | 20.354 |
| Average model loading time | 86.032 |
| Average time for initial prediction | 25.226 |
| Average time per plot, using model | 1.630 |
| Time per plot standard deviation, using model | 0.819 |

**Table 3.**
*Machine learning model vs. Scagnostics algorithm runtime evaluation, evaluated on 589 scatterplots [7].*

to the clumpy measurement. A large number of wrongly classified plots for clumpy scores suggest that we did not generate this kind of data properly, or the clumpy score is too sensitive [9].

### 5.4 ScagCNN run-time evaluation

There are two main advantages of using ScagCNN compared to the conventional Scagnostics algorithms. The first is that thanks to CNN's spatial invariant property, ScagCNN is robust to small changes in the underlying data. We refer interested readers to the original paper [9] for detailed analysis concerning Scagnostics's sensitivity and ScagCNN's robustness concerning small variants in the underlying data. This section discusses the second advantage of ScagCNN's. It reduces prediction time in case we need to predict the scores in batches.

**Table 3** summarizes the time performance on the test dataset with 589 scatterplots from the real-life datasets. The testing device is an iMac with 3 GHz 6-Core Intel Core i5, macOS Catalina Version 10.15.3, 8 GB of RAM. However, one may access the project's Github page to perform the runtime evaluation on his/her device. Notably, ScagCNN model has enormously reduced the computational time (20.019 *ms* vs. 1.630 *ms*). Furthermore, the conventional Scagnostics computation time depends on the number of bins resulting from the binning process. Thus, the conventional approach has a large standard deviation regarding the computation times.

In contrast, all inputs into ScagCNN are images of the same size, so the computation times are relatively more stable. However, it takes time to load the ScagCNN model and perform the first prediction. This initialization time is due to the WebGL shader compilation needed for the model [26]. Therefore, we should always use ScagCNN in batch to gain time performance. Further investigations reveal that ScagCNN prediction model achieves better performance by leveraging parallelisms (multi-cores) and using more memory space [7].

Finally, ScagCNN provides a web prototype for readers to evaluate the prediction results qualitatively. The source codes and result visualizations are available on the project's web page: https://idatavisualizationlab.github.io/V/ScagCNN.

## 6. Conclusions

This chapter introduces a recent Scagnostics implementation called ScagnosticsJS. Its implementation in JavaScript fosters the use of visual features on the web platform. Also, its extensions of Scagnostics to 3D and nD versions promote visual features that characterize the rapidly growing multivariate data. This chapter also describes Outliagnostics as an attempt to demonstrate the use of ScagnosticsJS 2D version on the web. This application uses a leave-one-out strategy and Scagnostics outlying score to detect individual data entry's outlying contribution to the overall set of points over time. Likewise, MTSAD is another application to illustrate the use of ScagnosticsJS nD version in finding the abnormalities in multivariate time series data. Finally, this chapter discusses a recent attempt to use convolutional neural networks to predict Scagnostics scores and classify scatterplots into their typical Scagnostics types. Its main purposes are to tackle the issues that the conventional Scagnostics algorithms involve time-consuming algorithms, and they are also sensitive to slight changes in the underlying data.

## Acknowledgements

This research is supported in part by the National Science Foundation under the I-Corps award number 2017018, grant CNS-1362134, OAC-1835892, and the IUCRC-CAC (Cloud and Autonomic Computing) Dell Inc. membership contribution.

## Author details

Vung Pham* and Tommy Dang
Texas Tech University, Lubbock, USA

*Address all correspondence to: vung.pham@ttu.edu

## References

[1] Wilkinson L, Anand A, Grossman R. Graph-theoretic scagnostics. Proceedings - IEEE Symposium on Information Visualization, INFO VIS. 2005:157–164.

[2] Lee W, Anushka A. Compute scagnostics - scatterplot diagnostics. rforge; 2018. Available from: https://www.rforge.net/scagnostics/.

[3] Josua K. Python binding to R scagnostics.. github; 2015. Available from: https: //github.com/nyuvis/scagnostics.

[4] Dang TN, Wilkinson L. ScagExplorer: Exploring scatterplots by their scagnostics. IEEE Pacific Visualization Symposium. 2014:73–80.

[5] Fu L. Implementation of Three-dimensional Scagnostics [master's thesis]. University of Waterloo; 2009.

[6] Pham V, Dang T. ScagnosticsJS: Extended Scatterplot Visual Features for the Web. In: Wilkie A, Banterle F, editors. Eurographics 2020 - Short Papers. The Eurographics Association; 2020..

[7] Pham V, Dang T. Outliagnostics: Visualizing Temporal Discrepancy in Outlying Signatures of Data Entries. In: 2019 IEEE Visualization in Data Science (VDS). Vancouver, BC, Canada, Canada: IEEE; 2019. p. 29–37.

[8] Pham V, Nguyen N, Li J, Hass J, Chen Y, Dang T. MTSAD: Multivariate Time Series Abnormality Detection and Visualization. In: 2019 IEEE International Conference on Big Data (Big Data). IEEE; 2019. p. 3267–3276.

[9] Pham V, Nguyen NV, Dang T. ScagCNN: Estimating Visual Characterizations of 2D Scatterplots via Convolution Neural Network. In: Proceedings of the 11th International Conference on Advances in Information Technology; 2020. p. 1–9.

[10] Wilkinson L, Wills G. Scagnostics distributions. Journal of Computational and Graphical Statistics. 2008;17(2): 473–491.

[11] Bostock M, Ogievetsky V, Heer J. D3 data-driven documents. IEEE transactions on visualization and computer graphics. 2011;17(12):2301–2309.

[12] Nürnberg R. Calculating the area and centroid of a polygon in 2d. URL: http://wwwf imperial ac uk/˜rn/centroid pdf. 2013.

[13] Dang TN, Wilkinson L. Transforming scagnostics to reveal hidden features. IEEE Transactions on Visualization and Computer Graphics. 2014;20(12):1624–1632.

[14] Sartipizadeh H, Vincent TL. Computing the approximate convex hull in high dimensions. arXiv preprint arXiv:160304422. 2016.

[15] WorldWebMath. N Dimensional Geometry; 1997. Last accessed 11 June 2019. Available from: http://web. mit. edu/wwmath/vectorc/ndim.html.

[16] Asuncion A, Newman DJ. UCI Machine Learning Repository. University of California; 2007. Available from: http://www.ics.uci. edu/~mlearn/MLRepository.html.

[17] Dang T. Visualizing Multidimensional Health Status of Data Centers. In: Bhatele A, Boehme D, Levine JA, Malony AD, Schulz M, editors. Programming and Performance Visualization Tools. Cham: Springer International Publishing; 2019. p. 273–283.

[18] Pham V, Nguyen N, Dang T. ContiMap: ContiMap: Continuous Heatmap for Large Time Series Data. In:

2020 IEEE Visualization in Data Science (VDS). Virtual Conference: IEEE; 2020.

[19] Hartigan JA. Clustering Algorithms. New York: John Wiley & Sons; 1975.

[20] Le DD, Pham V, Nguyen HN, Dang T. Visualization and Explainable Machine Learning for Efficient Manufacturing and System Operations. Smart and Sustainable Manufacturing Systems. 2019 Feb;3(2):20190029. Available from: https://doi.org/10.1520/ssms20190029.

[21] Krizhevsky A, Sutskever I, Hinton GE. Imagenet classification with deep convolutional neural networks. In: Advances in neural information processing systems; 2012. p. 1097–1105.

[22] Szegedy C, Liu W, Jia Y, Sermanet P, Reed S, Anguelov D, et al. Going deeper with convolutions. In: Proceedings of the IEEE conference on computer vision and pattern recognition; 2015. p. 1–9.

[23] Simonyan K, Zisserman A. Very deep convolutional networks for large-scale image recognition. arXiv preprint arXiv:14091556. 2014.

[24] Visual Geometry Group. Visual Geometry Group. http://www.robots.ox.ac.uk/~vgg/;. Accessed: 2019-11-4. http://www.robots.ox.ac.uk/~vgg/.

[25] Zhang A, Lipton ZC, Li M, Smola AJ. Dive into Deep Learning. https://d2l.ai; 2019. Accessed: 2019-11-4.

[26] TensorFlow. Tensorflow documentation. TensorFlow;. Accessed: 2019-12-24. https://github.com/tensorflow/tfjs/tree/master/tfjs-converter.

Chapter 2

# Visualizing the Impact of COVID-19 in the Mobility Dynamics - A Dashboard Framework for Decision Support in Smart Cities

*Nuno Alpalhão, Miguel de Castro Neto and Marcel Motta*


## Abstract

Being mobility one of the biggest challenge's cities face today, the COVID-19 pandemic reinforced this challenge and caused a deep structural change in the mobility of the multilayered dynamic framework of Smart Cities. The need to supply decision support systems to city authorities is higher than ever. Planning and managing mobility in Smart Cities has become more challenging, as the amount of information available and the pressure to enforce sustainable and secure policies increases, stakeholders require faster and more targeted actions. Dashboards are powerful tools that can be used in this context to provide, in an understandable manner, multidimensional information otherwise unavailable in classically static visualizations, as these tools offer a reliable foundation for decision support systems. This chapter goes through the required visualization techniques used to produce meaningful dashboards, to both showcase spatial and temporal trends in the context of mobility in Smart Cities following the COVID-19 pandemic. A general framework for analyzing mobility patterns is suggested by gathering methods and techniques recently developed in the literature.


**Keywords:** Smart Cities, mobility, COVID-19, decision support, dashboard, spatial and temporal trends

## 1. Introduction

The exponential growth and availability of data has opened the possibility of visualizing a city and all its layers in a previously unavailable smart way. We define the Smart Cities framework as an urban provider of several services clustered into different nonexclusive layers in a unified way [1], such clusters can be characterized as Mobility, Environment, Government, Economy, People and Living [2]. This digital transformation process cities face today is leading to a new reality where urban space is taking advantage of information and communication technologies and data science to answer present and future challenges, namely, to become more efficient in services and infrastructures management in order to deliver quality of life to the people who live, work or visit the city [3]. When developing a framework to support decisions in any of these layers one must always consider the amount of information available and its purpose. In this context it is natural to introduce a dashboard as a visualization tailored to give support to smart city agents, from managers to policy makers, in order to both understand and act on these complex matters [4] in a readily available manner and arranged on a single screen [5].

The global pandemic scenario caused by the Covid-19 pathogen raised many questions in terms of measuring and understanding its impact in multiple fronts (e.g. healthcare, economy, tourism). Addressing these questions became a critical task in tailoring and evaluating strategies to tackle the pandemic and minimize its effects, especially in the context of the Smart Cities. Naturally, a massive influx of dashboards started being developed, published and shared all over the internet by institutional agents, academia and industry. More often than not, these dashboards seem to lack well-defined guidelines in terms of their design choices and/or attempt to represent information in ways that are either unclear or dubious [6]. This diagnosis is one of the main drivers for writing this chapter and, hopefully, should provide an adequate reference guide for designing simple and insightful dashboards.

In this chapter we will provide a step-by-step dashboard design prototype applying structural guidelines [7] on how to deliver such an essential tool within the scope of mobility and the impact it suffered due to the Covid-19 pandemic in the city of Lisbon in Portugal. We will provide a broadening of the concepts required and encourage the reader to apply this methodology in any case within the framework of decision support in Smart Cities.

The proposed solution relies on a Javascript backend engine executed with Python programming language [8], in line with the state of the art. In comparison with other implementations, our dashboard improves the understanding of the impact of the Covid-19 pandemic by synthesizing visualization concepts and techniques gathered in the literature. This way, it is introduced a novel approach to better visualize mobility patterns in the context of Smart Cities.

## 2. State of the art

As defined in the literature, a dashboard is "a visual display of data used to monitor conditions and/or facilitate understanding". While there is not a clear definition in terms of its format, dashboards usually "combine different elements (e.g., charts, text, legends, filters, etc.) into a cohesive and coordinated whole that allows people to see and understand their data" [7].

In the field of mobility, it was identified a set of relevant dashboards that attempt to highlight trends and patterns by using clear visualizations and metrics, as shown in **Figures 1** and **2**:

Since the Covid-19 outbreak, Google started to release periodic reports [9] highlighting mobility changes at the city level, by using a category breakdown (e.g. Retail & recreation, Grocery & pharmacy, parks). While this report could arguably be described as a dashboard, it conveys information using data visualization tools and techniques to provide understanding over a certain event. In a simplistic manner, it defines a ratio metric to highlight the change in mobility caused by the policies and measures to tackle the Covid-19 outbreak. Additionally, it uses a line chart to analyze fluctuations over the previous weeks. Google also ensures the statistical significance of the information displayed by removing cities and categories which are not relevant or cannot be properly subject to analysis. On the other hand, the limited scope and lack of interactivity of this report could be pointed out as shortcomings and could be improved.

*DOI: http://dx.doi.org/10.5772/intechopen.96295*

**Figure 1.**
*Google's COVID-19 community mobility reports.*

**Figure 2.**
*C2SMART COVID-19 data dashboard.*

In **Figure 2**, an interactive dashboard is proposed by C2SMART [10] to visualize daily traffic conditions and report changes in commuting patterns in the city of New York since the Covid-19 outbreak. This dashboard consists of combining several data sources (i.e. taxi, subway, traffic jams and collisions) and techniques to provide a complete picture of the impact of the pandemic on transportation systems as it unfolds. Given the nature of data, we are once again presented with elements that can better display the spatial–temporal distribution of mobility: (a) a map visual displays the traffic flow for the main roads in the city; (b) a heatmap visual shows the year-over-year (YoY%) difference ratio in traffic volume; (c) a line-bar chart showing absolute and YoY% subway ridership data; (d) a line chart showing hourly travel times for a selected road; and (e) a bar chart showing the weekly reported crashes between 2019 and 2020. The depth of this dashboard poses as a great example for the usage of data visualization for analyzing the effects of Covid-19 in mobility. Nonetheless, it might present some challenges in terms of complexity when using the several available filters and interpreting the information in an adequate manner.

## 3. Mobility in the urban context

Mobility in the Smart City context is defined as the essence of contemporary cities, in other words it defines the interactions of all moving parts in the urban context with multiple and distinct information sources ranging from traffic sensors to telecommunications data. A concept such as this one can be used in the definition and planning of multiple services in the urban area, such as a major factor in a smart and sustainable urban planning, an economic proxy for socio-economic characterization a feature in implementing and monitoring security in cities and so on [11].

As we have mentioned in the beginning of this chapter, mobility is only one of the multiple layers in the framework of Smart Cities, from which new insights can be reached with the simultaneous use of other sources of information either explaining or being explained by it. That is, understanding mobility alongside other determinant factors, in the context for Smart Cities, allows to answer a variety of questions about how cities and their inhabitants interact, at a certain point in time and space. Intuitively, the answer is always dependent on the intent of the visualization itself, but good practices are transversal to all problems [12].

For the dashboard built later in this chapter we will represent mobility by encoding traffic data in the form of geometric objects to map and describe commuting events happening throughout the city of Lisbon between 2019 and 2020, for the months of September and October. Additionally, we will also be using telecommunications data in the same spatial and temporal range in the form of Origin Destination (OD) matrices [13] to understand and quantify how many people move between census tracts of the city. Due to the sensitive nature of the communications information, all the data related to the OD matrices used to feed the visualizations was artificially generated from the original source. As we will see in the sections bellow, the usage of these two data sources will allow us to visualize mobility patterns across the day before and after Covid-19.

## 4. Covid-19 impact scenario

As it is known the Covid-19 has been a prevalent pandemic that significantly changed the way people conduct their lives [14, 15], especially when it comes to mobility, as lock down and isolation measures restricted commuting and traveling to an unprecedented extent. With such a significant transformation it is natural that the need for governments to reinforce well-informed policies is higher than ever, with the ambition of moving to data-driven public policy making, especially in the urban areas where the incidence is more pronounced. From our definition of mobility, it comes naturally that a study on the impact of this pandemic is not only desirable but a necessity in the context of Smart Cities, particularly now that we begin to have sufficient historical data for the period of this pandemic. However, presenting and exploring data to effectively turn it into valuable information requires some considerations, such as: the definition of the observed event, the toolset and structure used in the report, the scope of the report, the target audience, and its main objectives.

Below we will start to define all the structural requirements along with a set of good practices for a successful and impactful dashboard.

### 4.1 Problem definition

In order to adjust city wide services and policies, such as public transport, security, traffic management and infrastructure development, city planners require not only the latest information regarding the current state of mobility but also how historically it changed in regard to the impact of the Covid-19 pandemic or, in other words, what is the new "normal" mobility patterns in the city of Lisbon?

We are expected to define and visualize the current context of the city's mobility as well as to be able to differentiate the impacts caused by the Covid-19 pandemic.

### 4.2 Visualization requirements

First and foremost, it is essential to outline a storytelling narrative behind your visualization to provide the proper context to the data, to highlight its relevant characteristics and, ultimately, to deliver insights.

Creating a visualization that can express the state of mobility at a given point in time, is the main objective, but having the city as a single entity is not enough to fully understand the dynamics of smaller partitions, such as neighborhoods and main roads. We identify three main requirements for a desirable dashboard: (a) a zoom in and out approach needs to be available to the user; (b) a clear temporal trend comparison is a must to understand the impact and the ways to administer the changes caused in the population's mobility; and (c) measures to understand if the current state of mobility is expected and if not, where it is deviating from the norm.

### 4.3 Scope

Will it be possible to use this dashboard for a different purpose, for example: can it be used to aid police officers in choosing patrol routes? By adding information would we lose on the main problem? These are the questions you should be making even before you start thinking visually, since they will surely affect the outcome and longevity of the dashboard itself.

### 4.4 Target audience

To whom the dashboard is aimed for should deeply impact its design, as you should not expect the end user to have the same degree of expertise in a subject as you do.

Given the importance of Smart Cities for policy makers and institutional agents, the main target of the proposed dashboard should be stakeholders whose responsibility is to tailor and gauge the conditions of the city's infrastructures, services and facilities made available to its inhabitants, to improve the management of resources, namely public transport, and to promote well-being. But taking into consideration the possible scope option mentioned in the section above, what if the dashboard was intended to police officers as well? Given that their line of work requires specific and fine-grained information to recognize which specific streets might need more attention, new layers of information would be needed to provide such an understanding.

## 5. A dashboard

### 5.1 Prerequisites

Before we can begin, we need to fully understand how we want to represent our problem set and the sources available for this purpose. For a descriptive visualization to successfully work it is required to operate under some well-founded assumptions.

For the problem at hand, we have available traffic jams, telecommunications OD matrices and geolocated layers such as metro stations, bus stops and road infrastructure. But how can we use them to represent such a broad term such as mobility? It is in this part of the chapter that we will start warning the reader that a single correct answer does not exist, different interpretations of the definition of the problem will naturally lead to different solutions, nevertheless, a substantiated approach that is able to effectively solve the problem is always of great value.

It is known that traffic jams are mainly caused by two intrinsically distinct factors [16]: (a) Daily commuting, a problem caused by a recurrent overload of the road system (mostly during week days) in both the home to work commuting (typically during the morning) and again in the analogous work to home commuting (typically during the afternoon), a problem most cities still struggle to solve; (b) A random event, road accidents or city wide events such as a sports match, can cause sporadic traffic jams. While the second point can bring some great insights on how to manage mobility given an external factor or event, it is the first point the allows us to consistently take a snapshot of the daily dynamics of mobility in a city, in order to characterize such a diverse structure as a city, recurrent events are always desirable.

With such a mindset we can consistently describe a factor of the city's mobility ecosystem on a daily or even weekly basis. To broaden our understanding of the mobility dynamics, using the OD matrices data, given a traffic jam, we can identify from which census tracts of the city people came from and went to.

Given the real time state of traffic, one cannot accurately identify the cause of a traffic jam, but if, for the same location, there is an occurrence of traffic jam in the morning and in the afternoon, and as it is known, commuting is recurrent, so with a daily, weekly, or even monthly overlay, we can start to visualize the typology of these jams and their implications.

## 5.2 Current state of mobility

The term current can be misleading, as the current state of mobility in the city is not solely described by the real-time state of traffic jams but by its past state as well. In the assumptions made, the recurrence of daily traffic jams can be identified as an overload or bottleneck on the road system caused by the commuting of people, nevertheless, to give the ability to the user to characterize and visualize a traffic jam, as commuting related or not, an encoding of the whole day needs to be present as well.

As one can see in **Figure 3**, the user has access to the daily state and degree, with a clean color encoding, green for home to work commutes and red for work to home, of traffic jams in the city and is simultaneously able to understand the extent of recurrence of these events. For a given temporal window (in the image set as "Daily"), an algorithm identifies and defines immediately, with a given degree of confidence, each commute as the reoccurrence of a traffic jam in the morning and afternoon in the same location, by analyzing the consistency of origin and destination census tracts between the morning and afternoon throughout the day or week, implicitly it already characterizes the mobility dynamics of a city by defining commuting of people as a daily occurrence through the same location, causing traffic jams. This visualization allows the user, in a very clear and succinct way, to spatially understand how mobility is being conducted, and the degree of severity of the overload in the road system. Allowing zoom in and out operations to clearly see the direction in which mobility is and historically where it has been going. Additionally, by choosing a concrete period and temporal aggregation on the top menu, the user can click the submit button to change the map, treemap and bar chart visualizations as desired.

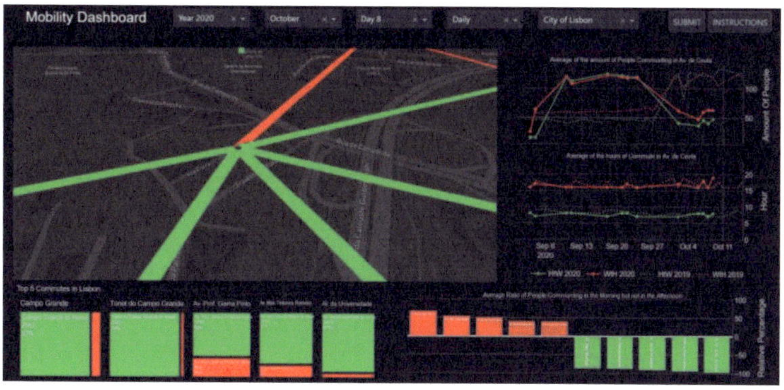


**Figure 3.**
*Dashboard prototype [17].*

By itself, the map is not enough to give an understanding of numerically how many people commute. How do we compare different commutes that are spatially separated? In the following visualizations we will focus on each commute as a bi-daily occurrence of a traffic jam on the same location and its origin and destination census tracts, as identified by our algorithm.

To understand the dimension and difference between different commutes, using the OD matrices we can calculate the total number of people that commuted from their homes to their work (HtW) and from work to their homes (WtH) for each traffic jam, under the assumption that it is caused by commuting. Naturally to guarantee, with a given degree of confidence, an average of the moving window for each type of commuted related jam can be made.

Having this information for the current state of mobility for that temporal window, a visualization can be created to compare different commutes and allow the user to understand the numeric difference of each unique commute and have a clear notion of the current state by comparing them.

It seemed appropriate to choose a treemap visualization [18] as shown in **Figure 4**, to allow the user not only to understand the number of people commuting but also to visualize the relative difference between them. In order to prevent an overcrowding in the visualization we chose to show only the top five commutes in terms of absolute number of commuters. Intuitively for this problem other types of visualizations could also have achieved the same level of effectiveness as a treemap visualization. For example, a pie chart could relay a better comparison within unique commutes but would not be so successful in displaying the comparison between distinct commutes. Although different choices could be made here, one should always prioritize on how the user should comprehend the data.

Now that a comparison of the number of commuting people between home and work has been made, given its spatial distribution, there is still the need to better understand how the difference in the number of people between the home and

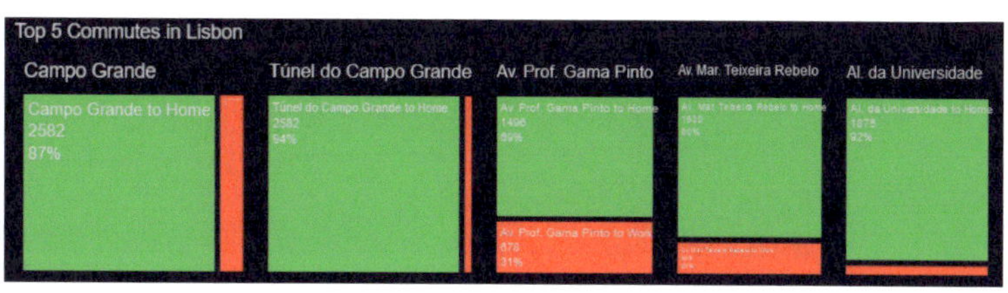


**Figure 4.**
*Treemap visualization.*

work commutes affect mobility. Intuitively, most people might commute to work daily in the same hour, participating in the same jam, but one cannot expect all of them to behave similarly, sharing again the same schedule when going back home. This translates in a consistent numeric difference that can be used to both characterize and differentiate different commutes.

Above in **Figure 5**, we have created a bar chart to visualize and compare such difference, each bar represents the relative difference of the number of people that commuted from home to work (HtW) by the number of people that commuted from work to home (WtH) for each commute, centered on the zero axis. In other words, the percentage delta of people that participated in a traffic jam in the morning for their morning commute and also participated in a traffic jam, in the same location in the afternoon. Again, to prevent overcrowding of the chart, only the top five highest and lowest ratios were shown. This visualization allows the user to understand how the main commutes in the city are differentiated spatially but also temporally from the morning to the afternoon, providing a new dimension in the comprehension of the city's dynamics.

### 5.3 Spatial–temporal data

The previous visualizations provide an understanding of the current state of mobility in a city given a temporal window by visualizing where and how

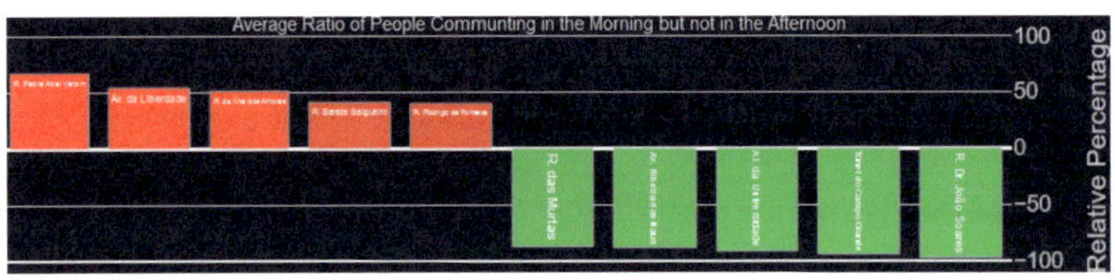


**Figure 5.**
*Bar chart visualization.*

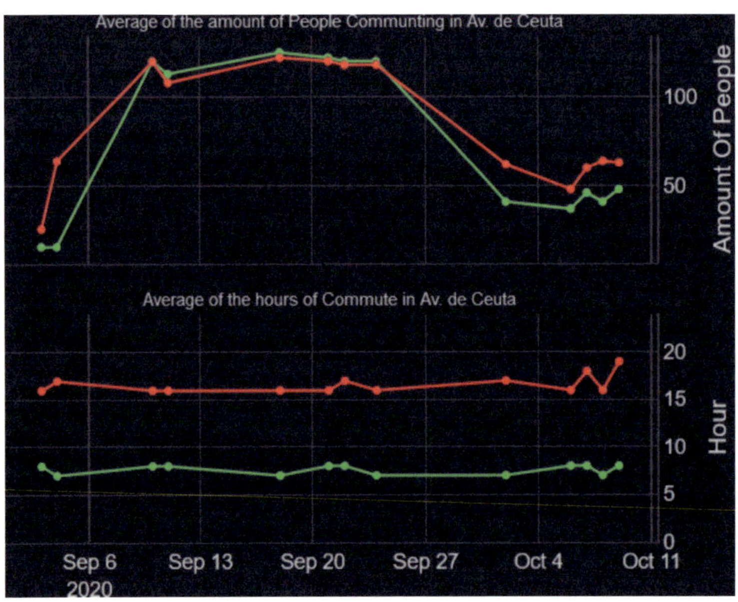


**Figure 6.**
*Line chart visualization.*

commutes are made. Nevertheless, to fully convey the natural change of how people commute throughout each yearly cycle, we need to analyze each individual jam location.

As we can see in **Figure 6**, by clicking in any specific point of the map, the user can access the full historical values of the amount of people that commuted in that location through the whole year and also, the average weekly hours in which the commute related jam occurred, this allows the user to grasp the existing seasonal trends to better discern the fundamental differences between the morning and afternoon commutes. By following the color encoding used in the map, using the greens for home to work transitions and reds for work to home, we create a cohesion in the dashboard that provides an ease of recognition for the user.

### 5.4 Measuring Covid-19 impact

To study the impact of such a prolonged event we first need to frame it temporally, as it is known the Covid-19 pandemic started spreading worldwide approximately in the month of February of 2020 and to the day it is still active. Naturally, a comparison to a period where the pandemic was not active is required, nevertheless due to seasonal trends the comparison needs to come from homologous months from different years.

As we can see in **Figure 7**, by using dotted lines, we can add the previous year's "Covid-19 free" information in the same line chart, where we can perfectly distinguish and compare the possible impact in mobility in terms of absolute values and hourly occurrence. The intention in the use of dotted lines in this visualization is to covey the attention of the user immediately, but also to set the previous year as a sort of target or something to be perceived as a goal, reflecting a hope in trying to achieve a "normal" mobility again.

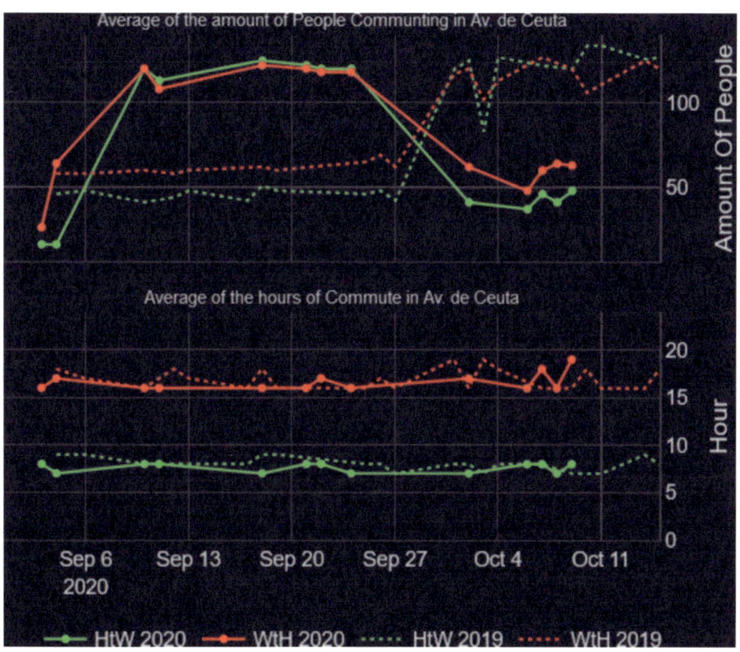


**Figure 7.**
*Visualization to offer a year-by-year comparison.*

## 6. Conclusion

In light of the Covid-19 outbreak and to support a data-driven urban management strategy, the need for data analytics tools is in the forefront of policymaking. In practice, this approach proves to be an interesting path for presenting and exploring shifts in mobility patterns, while being robust enough for answering transversal questions regarding the cities and its dynamics with the ambition of supporting data-driven policy making in pandemic situations.

This chapter presented and discussed dashboard solutions designed to analyze the impact of Covid-19 in mobility and proposed a framework for data visualization in the context of the Smart Cities using dashboard techniques and following a set of clear goals and good practices to reach them. Moreover, a dashboard prototype was proposed for visualizing changes and shifts in the local dynamics given historical data obtained before and after the pandemic outbreak in 2020.

Our findings seem to fit the current literature, as one can see, in **Figure 4** the use of a treemap visualization can be highly scalable while providing an efficient use of space in the dashboard itself [18], defining the problem, scope and target audience is essential prior to the development of any visualization [7, 12] and understanding Smart Cities as multilayered entities is indispensable in order to provide any kind of meaningful insights [1, 3], especially when dealing with mobility [11, 16].

Furthermore, the visual components proposed attempt to look at mobility data from multiple perspectives, in line with previous works; for instance, time can be visualized as a continuous dimension in a line chart to represent a historical series, or it could also be visualized as an ordinal variable in a bar chart to represent distributions across days of the week [7]. The proposed components should also allow to compare different points in time and space [7], in order to identify seasonal trends and/or spatial concentration, which could be achieved by line charts, maps and treemaps. The ideas herein discussed and the proposed guidelines are a small contribution to consolidate the application of dashboards in the field of the Smart Cities and, hopefully, this chapter could be used as inspiration for authors and contributors for further development in this field of study.

## Author details

Nuno Alpalhão, Miguel de Castro Neto* and Marcel Motta
NOVA Information Management School (NOVA IMS), Universidade Nova de Lisboa, Lisboa, Portugal

*Address all correspondence to: mneto@novaims.unl.pt

## References

[1] Anthopoulos, Leonidas. Understanding Smart Cities - A tool for Smart Government or an Industrial Trick? Springer. 2017. DOI:10.1007/978-3-319-57015-0

[2] Giffinger, Rudolf & Fertner, Christian & Kramar, Hans & Kalasek, Robert & Milanović, Nataša & Meijers, Evert. 2007. Smart cities - Ranking of European medium-sized cities.

[3] Neto, M. D. C. & Cartaxo, T. D. M. How Smart Is Your City?: Technological Innovation, Ethics and Inclusiveness. Ferreira, M. I. A. (ed.). Springer Science and Business Media B.V. 1 Jan 2021. Vol. 98. p. 59-73 15 p. (Intelligent Systems, Control and Automation: Science and Engineering; vol. 98).

[4] Pomerol, Jean-Charles & Adam, Frédéric. Understanding Human Decision Making – A Fundamental Step Towards Effective Intelligent Decision Support. 2008. DOI:10.1007/978-3-540-76829-6_1.

[5] Stephen Few. *Information Dashboard Design: The Effective Visual Communication of Data*. O'Reilly Media, Inc. 2006

[6] MIT Technology Review, The best, and the worst, of the coronavirus dashboards. Accessed: 12 December 2020. Available from: https://www.technologyreview.com/2020/03/06/905436/best-worst-coronavirus-dashboards/.

[7] Steve Wexler, Jeffrey Shaffer, Andy Cotgreave. The Big Book of Dashboards: Visualizing Your Data Using Real-World Business Scenarios. 1st ed. Wiley Publishing. ISBN: 1119282713

[8] Van Rossum G, Drake FL. Python 3 Reference Manual. Scotts Valley, CA: CreateSpace; 2009.

[9] Google, COVID-19 Community Mobility Reports. Accessed: 12 December 2020. Available from: https://www.google.com/covid19/mobility/

[10] Zuo, F., Wang, J., Gao, J., Ozbay, K., Ban, X. J., Shen, Y., ... & Iyer, S. (2020). An interactive data visualization and analytics tool to evaluate mobility and sociability trends during Covid-19. arXiv preprint arXiv:2006.14882.

[11] Arce-Ruiz, Rosa & Baucells, Neus & Moreno Alonso, Concepcion. (2016). Smart Mobility in Smart Cities. 10.4995/CIT2016.2016.3485.

[12] Sarikaya, Alper & Correll, Michael & Bartram, Lyn & Tory, Melanie & Fisher, Danyel. (2018). What Do We Talk About When We Talk About Dashboards?. IEEE Transactions on Visualization and Computer Graphics. PP. 1-1. 10.1109/TVCG.2018.2864903.

[13] Md. Shahadat Iqbal, Charisma F. Choudhury, Pu Wang, Marta C. González, Development of origin–destination matrices using mobile phone call data, Transportation Research Part C: Emerging Technologies, Volume 40, 2014, Pages 63-74, ISSN 0968-090X, https://doi.org/10.1016/j.trc.2014.01.002.

[14] Mahmudur Rahman Fatmi. COVID-19 impact on urban mobility. Journal of Urban Management. Volume 9, Issue 3. 2020. ISSN:2226-5856. DOI:10.1016/j.jum.2020.08.002.

[15] Gilbert, Marius et al. Preparedness and vulnerability of African countries against importations of COVID-19: a modelling study. The Lancet, Volume 395, Issue 10227, 871-877

[16] Tanzina Afrin and Nita Yodo. A Survey of Road Traffic Congestion Measures towards a Sustainable and Resilient Transportation System.

Sustainability. 2020,12, 4660; DOI:10.3390/su12114660

[17] Dashboard Prototype, Mobility Dashboard. Accessed: 26 January 2021. Available from: https://mobility-dashboard-chapter.herokuapp.com/

[18] Lim Kian Long, Lim Chien Hui, Gim Yeong Fook, Wan Mohd Nazmee Wan Zainon, A Study on the Effectiveness of Tree-Maps as Tree Visualization Techniques, Procedia Computer Science, Volume 124, 2017, Pages 108-115, ISSN 1877-0509, https://doi.org/10.1016/j.procs.2017.12.136.

Section 2

# Digital Twins

Chapter 3

# 3D Point Cloud-Based Tree Canopy Visualization for a Smart Deployment of Mobile Communication Systems

*Yunus Egi and Engin Eyceyurt*


## Abstract

Mobile communication is one of the most important parameters of smart cities in terms of maintaining connectivity and interaction between humans and smart systems. However, In the deployment process of Mobile Communication Systems (MCS), Radio Frequency (RF) engineers use location depended empirical Signal Strength Path Loss (SSPL) models ending up with poor signal strength and slow data connection. This is due to the fact that empirical propagation models usually are restrained by the environment and do not implement state of the art technologies, including Unmanned Aerial Vehicles (UAV), Light Detection and Ranging (LiDAR), Image Processing, and Machine Learning to increase efficiency. Terrains involving buildings, hills, trees, mountains, and human-made structures are considered irregular terrains by telecommunication engineers. Irregular terrains, specifically trees, significantly affect MCS's efficiency because of their complex pattern resulting in erroneous signal fading via multi-path reflection and absorption. Therefore, a virtual 3D environment is required to extract the required 3D terrain pattern and elevation data from the environment. Once this data is processed in the machine learning algorithm, an adaptive propagation model can be formed and can significantly improve SSPL prediction accuracy for MCS. This chapter presents 3D point cloud visualization via sensor fusion and 2D image color classification techniques, which lead to a novel propagation model for the smart deployment of MCS. The proposed system's main contribution is to develop an intelligent environment that eliminates limitations and minimizes related signal fading prediction errors. In addition, having better connectivity and efficiency will resolve the communication problem of smart cities. The chapter also provides a case study that significantly outperforms other empirical models with an accuracy of 95.4%.


**Keywords:** point cloud, image processing, direct geo-referencing, machine learning, radio networks, signal strength path loss

## 1. Introduction

In recent years, the number of the mobile subscriber has reached 5.11 billion worldwide with a 2% increase [1]. This is due to the fact that available technologies such as LTE and 5G are in use in smart cities to communicate and share the data through video conferencing, online shopping, working from home, smart transportation and etc. In addition, Covid-19 lockdowns cause 70% increase in internet usage and 12% increase in streaming [2]. Accessibility of higher throughput transfers in uplink and downlink led to accelerating the digitization process of new technological advancements including Internet of Things (IoT), cloud computing, big data analysis, and machine learning [3]. The digitization process is crucial in terms of creating smart cities. As the usage of LTE and 5G technologies increases in smart cities, the low-cost broadcast systems designed for different terrains become more critical. Lands containing mountains, hills, vegetation, and human-made structures are considered irregular terrains in telecommunication due to their complex structure and surface pattern. However, vegetation has more impact on wireless communication since leaves and branches cause faster absorption of signal strength through multi-path reflection [4, 5]. Empirical propagation models such as Free space, Log-Normal (LN), and Cost-231 Hata models are limited by the environment selection which may lead to inaccurate results in SSPL predictions [6, 7]. In order to avoid faulty results, it is essential to take into account both the elevation and the terrain shapes in SSPL calculations. For instance, in **Figure 1**, ignoring the tree canopies falling into the Fresnel zone between the transmitter and the receiver affects the performance of the transmitted signal and causes false path loss estimation as the main signal has interference from the tree canopy. It is not possible to physically distinguish between vegetation and the rest of the surrounding environment in practical applications. When equipped with a geo-referenced satellite image and a corresponding geo-referenced 3D point cloud, it becomes possible to create virtual twin environment and extract the features of the environment to predict SSPL as seen in **Figure 2**. In airborne LiDAR, Geographical Mobile Mapping System (Geo-MMS), the GPS and IMU provide the exact location and orientation of airborne LiDAR [8]. The LIDAR sends laser pulses to objects on the earth and collects the reflected pulses from the environment. The distances are found by calculating the time delay ($\mu$) between the transmitted and received laser pulses [9]. The obtained geo-referenced data points form a 3D point cloud that will give the surface's property, such as the height and width of the obstacles. However, since these features will not be sufficient to classify trees from the environment, 2D satellite imagery will be used. The 2D satellite image has a colorful and classifiable character. We can use this feature to extract trees from the 3D image by classifying them. This process is illustrated in **Figure 3**.

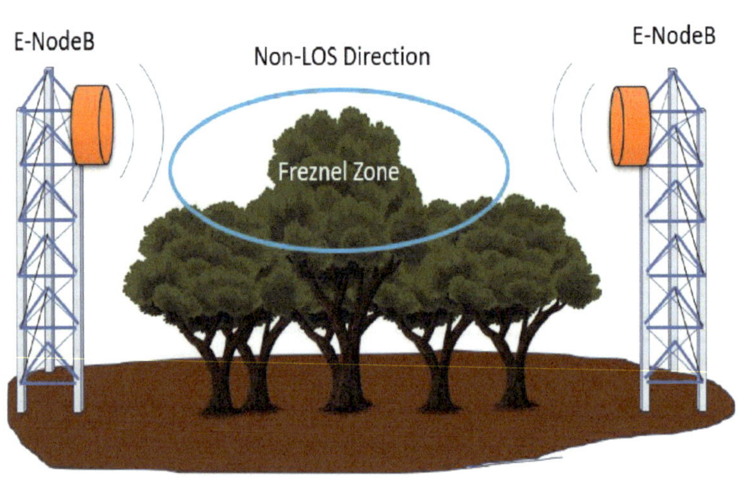


**Figure 1.**
*Illustration of non line-of-sight (NLOS) propagation.*

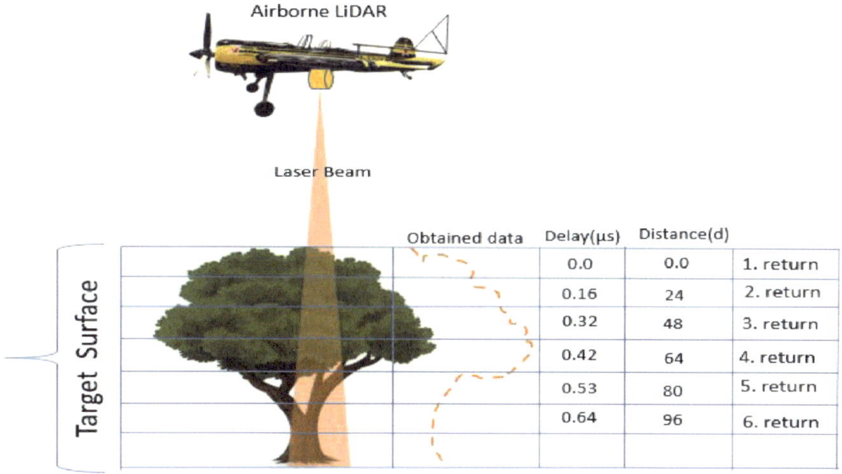


**Figure 2.**
*Obtaining 3D point cloud using airborne LiDAR.*

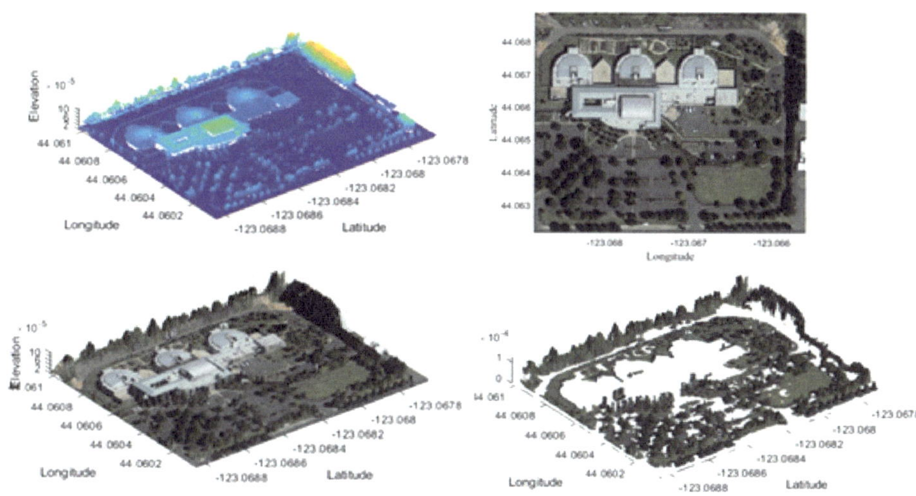


**Figure 3.**
*3D Point cloud vegetation classification process.*

Machine learning determines the learning patterns and builds a general decision algorithm by training some portion of data and is one of the most important parameters of smart cities. In this research, Artificial Neural Network (ANN) which is one of the most widely used machine learning algorithms will be used to build an adaptive SSPL model by using features obtained from 3D virtual environment. ANN consists of an input layer, hidden layer, and output layer [10]. A simplified illustration of an ANN is represented in **Figure 4**. A simple ANN with only one hidden unit is shown in **Figure 4**. In the input layer, the data $(x_1, x_2, .., x_n)$ and randomly initiated weights $(\theta_1, \theta_2, .., \theta_n)$ are multiplied and transmitted to the hidden layer, also known as the activation layer. In the hidden layer, a bias unit is added to the sum of the processed data, and the result goes through an activation function. A non-linear sigmoid function is usually used as an activation function and assigns 1 or 0 based on a threshold. In the next step, a gradient descent algorithm is used to adjust the connection weights between neurons. Gradient descent calculates errors between predicted and real values and finds the neuron

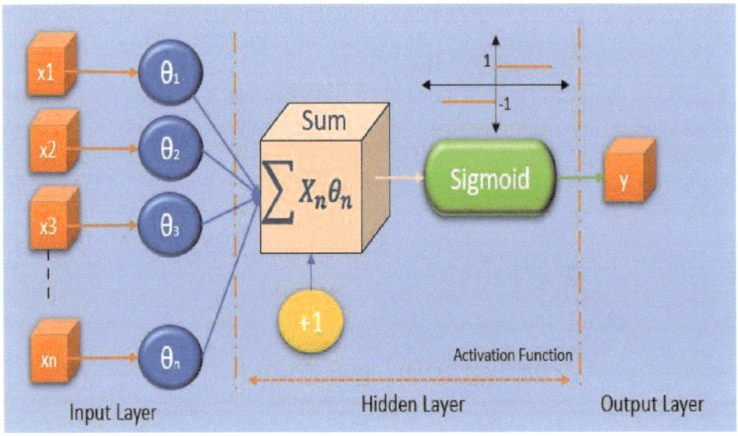


**Figure 4.**
*Simple Presentation of a Perceptron.*

weights that minimize the error by iteratively taking the gradients. This section will include an example using ANN with multiple hidden layers to estimate the SSPL of tree canopies. In this study, a powerful SSPL model has been created by using LiDAR and machine learning to solve MCS's aforementioned problems. The proposed model, having a combination of ANN and 3D point cloud, applies to large- and small-scale applications.

## 2. Related work

Many comprehensive studies have been conducted on Wireless Sensor Networks due to the continuous growth of mobile communication in smart cities. The studies mostly focus on power consumption and system accuracy [11]. In each study, different environments were tested over various aspects with image classification. However, image classification usage is not limited to a specific area but can be seen in almost all areas of studies. Juheon et al. performed an extensive study on the classification of individual tree species with LiDAR and deep learning [12]. Similarly, Torabzadeh et al. also studied tree species classification in forests with a combination of spectroscopy and airborne LiDAR [13] while Hartling et al. focused on urban tree species recognition and classification [14]. Image classification is a complex technique that finds space in many research areas. Image processing plays a crucial role in path loss analysis. In his study, Thrane aims to find the impact of buildings and multi-path propagation path loss of predefined signals [15]. He collects the signal attenuation measurements between the transmitter and receiver located in many different positions. After a 2D satellite image of the measurement area is obtained via Google Maps and rotated versions of images are prepared, the path loss effects of buildings are estimated with image classification and deep learning techniques. He states that 1 dB to 4.7 dB improvement factor in path loss prediction with his path loss model compared to empirical models. Likewise, the research conducted by Ahmadien et al. demonstrates a path loss model with K-mean clustering, deep learning techniques, and 3D images that converted from 2D satellite images via various simulation software [16]. Although 2D images are insufficient to create 3D images in many cases, he proposed a simulation-based path loss model in his study with limited parameters. Gracchi et al. uses a 3D point cloud taken by LiDAR to optimize WSN installation. High-resolution 3D point cloud data is analyzed to find a clear line of sight. She has validated her simulated 3D version of the model with the LiDAR-based 3D point cloud visualization [17]. The next section will provide 3D image segmentation using geo-referenced satellite images.

## 3. Visualization of Trees Using Image Classification and 3D point Cloud

### 3.1 Direct geo-referencing and multi-sensor fusion

Direct geo-referencing is a highly efficient and accurate technique used to determine the location and direction of a Geo-MMS. Measurements of external orientation parameters such as altitude, orientation angles, and distances are used in geo-referencing. A cameraless illustration of the Geo-MMS system is shown in **Figure 5**.

Downward directed LiDAR under the plane takes rotational scans while IMU and GPS take separate measurements. The direct geo-referencing of a ground vector on the surface is computed by Eq. (1) [18].

$$\vec{X}_k^a = \vec{X}_b^a + R_b^a\left[R_L^b R_k^L \vec{x}_k^L + \Delta\vec{x}_b^L\right]; \tag{1}$$

Where: $\vec{X}_k^a$: The ground vector ($a^{th}$ frame), $R_L^a$ and $R_L^b$: Rotation matrix and boresight rotation, $R_k^L(\theta, \eta)$: The LiDAR mirror rotation, $\vec{x}_k^L$ and $\Delta\vec{x}_b^L$: Slant range of the LiDAR and boresight-shifts. The $R_k^L(\theta, \eta)$ is the function of the angle $(\theta)$ between ground and target. The angle $(\eta)$ between ground and laser's X direction is calculated as follows.

$$R_k^L(\theta, \eta) = \begin{bmatrix} \cos\theta & -\sin\theta\sin\eta & -\sin\theta cos\eta \\ 0 & \cos\eta & -\sin\eta \\ \sin\theta & -\cos\theta\sin\eta & \cos\theta\sin\eta \end{bmatrix} \tag{2}$$

Kalman filter supported Inertial Navigation System (INS) geo-locates the data received by LiDAR and IMU by GPS data. A sequential adjustment is required since

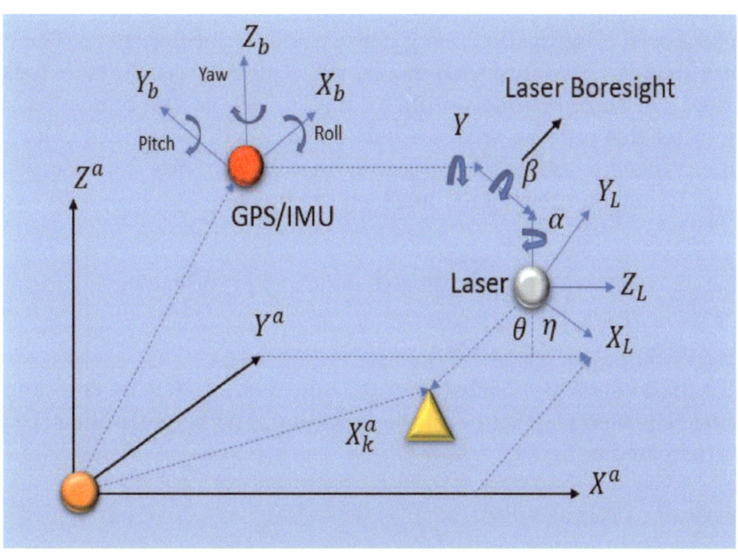


**Figure 5.**
*Obtaining 3D point cloud using airborne LiDAR.*

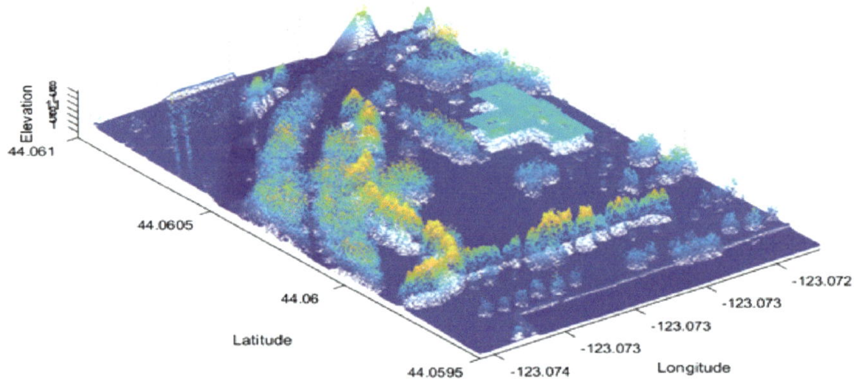


**Figure 6.**
*2D satellite image of Florida Tech neighborhood.*

each sensor operates in different frequencies. After synchronization is ensured and direct geo-referencing is complete, geo-referenced IMU and LiDAR data points are combined to visualize a 3D point cloud. An example of a 3D point cloud that belongs to the Lane County Mental Health service neighborhood located in Oregon is illustrated in **Figure 6** [19]. The LiDAR used in this example is operating at 70 kHz scanning speed and 1 cm resolution. The data is interpolated to fill the unmeasured spots and minimize the measurement error via Natural Neighbor Interpolation (NNI) [20]. It is also seen that the LiDAR data is geo-referenced in the x and y direction as longitude and latitude, respectively.

The corresponding 2D satellite image is also required for 2D/3D image fusion. The Google Map is utilized to extract the required geo-referenced 2D satellite image as represented in **Figure 7**.

### 3.2 Color classification for visualization of trees

Image classification is implemented to visualize necessary and informative properties using various methods through image processing. In this section, we apply a color-based classification using density and LAB color space. The images in RGB format are converted into LAB images since they are not suitable for digital manipulation. The three channels of LAB color space L, a, and b must be evaluated separately. After this process, with the help of Eq. (3), the intended color on the image will be picked to create a binary mask that will provide the average color of each channel falling on the selected mask on the image [21].

$$\mu = \left( \frac{1}{m \times n} \sum (m, n) P_{mask}(m, n) \right) \times Ones_{m \times n} \tag{3}$$

Find :$\mu_{maskL}, \mu_{maska}, \mu_{maskb}$

Relevant masks such as $mumask$, $mu_{maska}$ and $mu_{maskb}$ will be computed for each channel. The next step is to find the difference ($\Delta$) between the masks and the channel as indicated in Eq.(4).

$$\Delta L_{M \times N} = L - \mu_{maskL} \quad , \Delta L_{M \times N} = a - \mu_{maska} \quad , \Delta L_{M \times N} = b - \mu_{maskb} \tag{4}$$

**Figure 7.**
*2D satellite image of Lane County Mental Health service.*

Since the masks alone represent only the drawn area, calculating the Euclidean distance for all three channels with the formula $\Delta E_{M\times N}$ will reveal the color values closest to the masked part of the image as seen in Eq. (5).

$$\Delta E_{M\times N} = \sqrt{(\Delta L)^2 + (\Delta a)^2 + (\Delta b)^2} \tag{5}$$

To obtain an efficient classification, the color estimation ($\Delta E_M$) should fall in the 95% Confidence Interval (CI). This is essential since sharp color gradations within the area without tolerance can cause some areas to disappear. Thus, 95% CI should be applied to results by adding $3\sigma$ as seen in Eq.(6).

$$CI = \mu(\Delta E_{masked}) + 3\sigma(\Delta E_{Masked}) \tag{6}$$

Next, CI is applied to the $\Delta E$ to test $\Delta E <= CI$ values. If the condition is correct, logic one is assigned to this value. If not, logic 0 is assigned. The logical image containing zeros and ones is represented in **Figure 8a**. After implementing the logical image to the original image, the classified trees are obtained on a 2D satellite image. The classified image is demonstrated in **Figure 8b**. Since we aim to classify trees on the 3D point cloud, we can use that classified image to filter out undesired parts of the 3D point cloud other than trees, see **Figure 8c**. The results are shown in **Figure 9**.

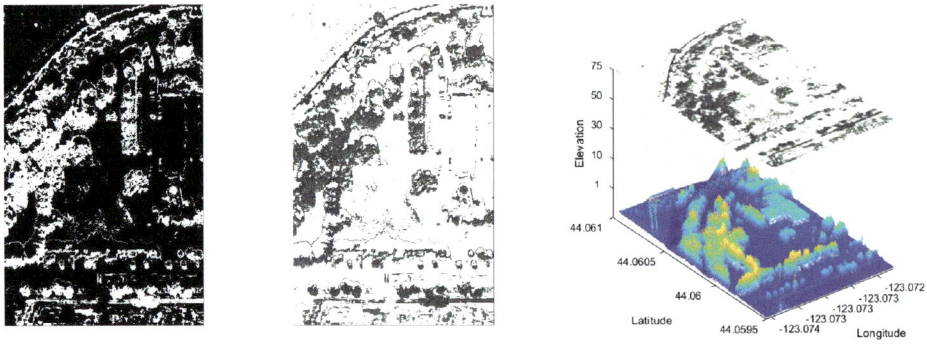


(a) Binary Filter (b) Classified Images (c) Filtering 3D image using 2D image.

**Figure 8.**
*3D Filtering process of tree canopies.*

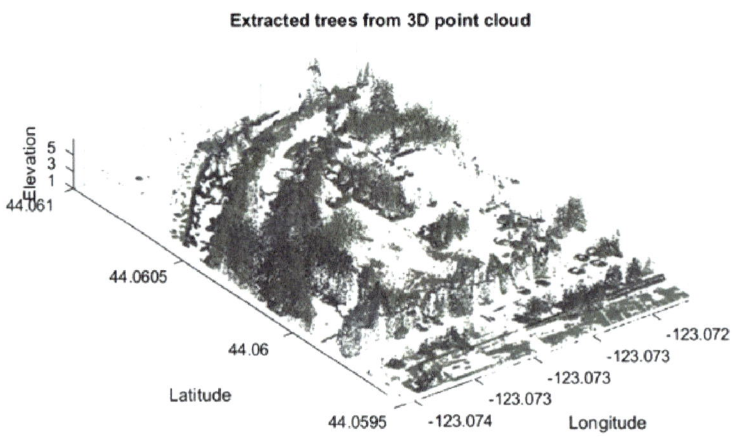


**Figure 9.**
*Obtained extracted vegetation and trees from the environment.*

## 4. Empirical models and proposed smart deployment technique

The signal's strength weakens when the signal encounters obstacles and loses energy due to multi-path reflection and absorption. The sum of the power loss and the signal path is called the signal strength path loss, which is decisive when deploying MCS. Therefore, many researchers have created path loss models such as Free-Space, Log-Normal, and Cost231-Hata models. Since each model emerges from experiments at a specific location, it has unique approaches specific to that location. In this research, we want to create a model supported by artificial intelligence and can fit in any location by getting out of location-oriented models that can be considered a disadvantage. All models, including the model to be obtained, were compared with each other. This model has been validated using the Mean Absolute Percent Error (MAPE).

### 4.1 Free-Space Path Loss Model

The free space path loss (FSPL) model determines the attenuation between the transmitter and the receiver in an unobstructed path. This phenomenon is indicated by the Friis transmission formula, as indicated in Eq.(7) [22].

$$FSPL = 10\log\left(\frac{P_t}{P_r}\right) = 10\log\left(\frac{(4\pi d)^2}{\lambda^2 G_t G_r}\right) \tag{7}$$

Where: $\lambda$: Wavelength, $G_t$: Gain of the transmitting antenna, $G_r$: Gain of the receiving antenna, $d$: Separation between transmitter and receiver.

### 4.2 Log-Normal Shadowing Path Loss Model

Log-normal shadowing (LNS) is the extended version of the Friis formula, which includes obstacles to the free space. It is a frequently used model for long-range propagation [23]. Because of the shadowing effect, the LNS model comprises Additive white Gaussian Noise (AWGN) represented as $X_\sigma$ [24]. The LNS model is demonstrated in Eq. (8).

$$PL_{LNS}[dB] = PL(d_0) + 10\eta\log\left(\frac{d}{d_0}\right) + X_\sigma \tag{8}$$

Where: $PL(d_0)$: The path loss at $d_0$, $\eta$: Path loss exponent, $X_\sigma$: $N(0,\sigma)$ Normal distribution with zero mean. LNS model has the following environments and Path loss exponents, as shown in **Table 1**.

### 4.3 Cost231-Hata model

Cost 231-Hata model is an SSPL model that takes the Okumura Hata model to a more diverse frequency range (1500–2000 MHz). This model can be used mainly for urban, suburbans, and open areas. The Cost 231-Hata model is indicated in Eq. (9) [25].

$$\begin{aligned} PL_{C-231}[dB] &= 46.3 + 33.9\log(f) - 13.82\log(h_B) \\ &\quad - a(h_R,f) + (44.9 - 6.55\log(h_B))\log(d) + C \end{aligned} \tag{9}$$

$$\text{Where } a(h_R,f) = (1.1\log(f) - 0.7)h_R - (1.56\log(f) - 0.8)$$

$$C = \begin{cases} 0\,\text{dB for Suburban areas} \\ 3\,\text{dB for Urban areas} \end{cases}$$

### 4.4 Proposed smart MCS deployment Technique

To design an intelligent broadcasting model for the deployment of MCS, the impact of vegetation must carefully be defined in the LOS direction. Towards LOS, the signal will be attenuated by transmission across vegetation due to reflections and absorption. Therefore, trees are measured as a highly complex obstruction in the environment from a telecommunication perspective. This phenomenon is expressed in **Figure 10**.

| Environment | Path Loss Exponent ($\eta$) |
|---|---|
| Free Space | 2 |
| Urban area(Shadowed) | 3–5 |
| Inside building(LOS) | 1.6–1.8 |

**Table 1.**
*Predefined Path loss exponent ($\eta$) for different environments.*

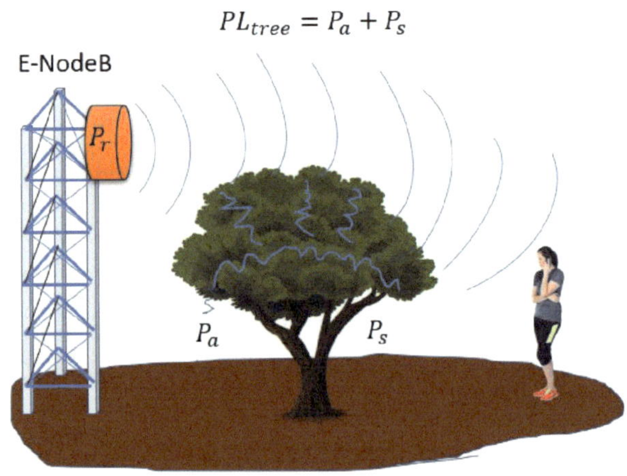


**Figure 10.**
*Illustration of tree canopy path loss system.*

This model is based on signal loss occurring on tree canopies ($PL_{tree}$) through scattering and absorption. In addition, considering $P_a$ and $P_s$ as the tree loss factor $PL_{tree}$, the corresponding signal strength loss is calculated as follows [26].

$$SSPL_{tree} = FSPL + PL_{tree} \tag{10}$$

In the literature, analysis of tree SSPL is a challenging ongoing research question. Even though the research numbers are going up, there are still a few consistent results due to the complex structure of the surveying area. This section extracts the features by means of image color classification and uses those features to reveal the required tree canopy path loss and add its effect to the FSPL through the ML algorithm. Since absorption and scattering are positively associated with the tree canopy's height and width, our algorithm will use these properties and estimate the required signal strength path losses.

### *4.4.1 Experiment and Model Presentation of Tree Canopy Path loss: case study*

In this section, a experiment conducted by Egi et al. [18]. will be evaluated. From his study, the necessary training data were collected using a 40 m Mini Handheld Digital Laser Range Finder, LG G5 mobile phone. The application used for this experiment is a network activity application called Network Cell Info (by Wilysis). The phone is fixed in the LOS direction of the transmitter. To calculate $PL_{tree}$, the data is taken from the front and back sides of the tree canopy and subtracted from each other. This difference corresponds to the $PL_{tree}$ caused by the scattering and absorption. This procedure is replicated in various places for different sizes of tree canopies. Since the data obtained are raw, feature normalization is applied to cause the gradient descent to converge faster. Mathematically explained, normalization is to subtract each feature's mean value from each item of the corresponding feature and scale the feature according to its standard deviation. Normalized data are used as input in our proposed model, as seen in **Figure 11**.

In this part, $x_1(height)$ and $x_2(width)$ are normalized and fed to the ANN. As the data pass through the perceptrons, they are multiplied by randomly initialized coefficients, called weights, $(\theta^l)$. At each layer, bias units (+1) are added to the data, which contribute to ANN's outcome by modifying the activation functions [27]. The parameters $x_1, x_2$ and, +1 with 200 elements are the input layers. The second layer is

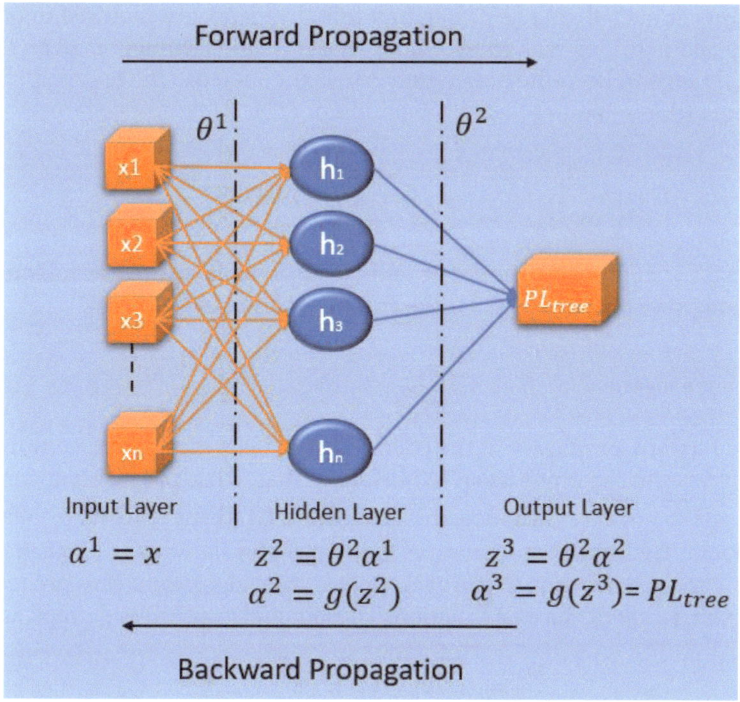


**Figure 11.**
*Proposed ANN presentation.*

the hidden layer consisting of 30 hidden units. The third layer is the output layer, which makes decisions based on the height and width taken from trees.

### *4.4.2 Calculating Cost Function Using Forward-propagation*

In ANN, the data travels from the input layer to the output layer to make predictions. Since the propagation is only one direction towards forward, it is called forward-propagation. The data are exposed to weights and non-linear sigmoid functions throughout propagation to add non-linearity into the estimation model. A regularization parameter can also be used to increase the prediction accuracy [27]. After the prediction through forward-propagation, the cost is computed to measure the performance of the ANN model. The cost is calculated by cost the function which determines the error between real and predicted values. The cost function may differ based on the purpose of ANN. In this chapter, we use the logistic regression cost function, as seen in Eq. (11).

$$J(\theta) = \frac{1}{m}\sum_{i=1}^{m}\sum_{k=1}^{K=3}\left[-y_k^{(i)}\log\left(\left(PL_{tree}\left(x^{(i)}\right)\right)_k\right) - \left(1-y_k^{(i)}\right)\log\left(1-\left(PL_{tree}(x)^{(i)}\right)_k\right)\right] + \frac{\lambda}{2m}\left[\sum_{j=1}^{30}\sum_{k=1}^{2}\left(\theta_{j,k}^{(1)}\right)^2 + \sum_{j=1}^{3}\sum_{k=1}^{30}\left(\theta_{j,k}^{(1)}\right)^2\right] \tag{11}$$

where: $PL_{tree}\left(x^{(i)}\right)$ is the last activation function, K and m are the number possible outcomes and number of labels respectively, y is the observed outcome, $\theta$'s are the weights and $\lambda$ is regularization parameter. The regulation parameter, $\lambda$, is used to prevent over-fitting [26]. Random initial weights are required to break the symmetry and to utilize each hidden unit. Initial weights should be given in the $[-\varepsilon_{init}, \varepsilon_{init}]$ range to keep the parameters small and increase the learning efficiency. The formula for the required $\varepsilon_{init}$ is given in Eq. (12).

$$\varepsilon_{init} = \frac{\sqrt{6}}{\sqrt{L_{in} + L_{out}}} \tag{12}$$

Where: $L_{in}$ *and* $L_{out}$ are number of units in adjacent layers. After forward-propagation process with $\lambda = 0.01$, the cost, $J$, is found as 2.052.

### 4.4.3 Back-propagation

Unlike forward-propagation, the back propagation propagates backward from the output layer to the input layer. While doing that, back-propagation computes gradients $(g'(z)^l)$ in every step towards backward. Gradient reveals $(\delta_j)$ changes in hidden layers. The subscript $J$ indicates the number of iterations and changes regularly with each iteration in the back-propagation algorithm. This process is used for the optimization of the cost function. The sigmoid gradient is defined as follows:

$$sigmoid(z) = g(z) = \frac{1}{1 + e^{-z}} \tag{13}$$

$$g'(z) = \frac{d}{dz} g(z) = g(z)(1 - g(z)) \tag{14}$$

In the model, $(delta)$ errors given in Eq. (15) cause deviation and must be calculated in every layer [26].

$$\delta_j^{(3)} = \alpha_j^{(3)} - y_i \rightarrow \quad Output \quad layer$$

$$\delta_j^{(2)} = \left(\theta^2\right)^T \delta^{(3)} .* g'\left(z^2\right) \rightarrow \quad Hidden \tag{15}$$

$$\Delta^l = \Delta^l + \delta^{(l+1)} \left(\alpha^{(l)}\right)^T \rightarrow \quad Adding$$

The outcome of $\Delta^l$ is an unreqularized gradient of the ANN cost function, which should be divided by $m$ (total number of samples). During the process, it updates the $\theta$ value for all j's simultaneously. After training the data with a hundred forward and back-propagation, the cost was reduced from 2.052 to 0.636, resulting in an accuracy of 94.5% in signal strength estimation. Since the predicted accuracy is reached with our ANN algorithm, we apply our algorithm in the direction of LOS to the detected trees. Tree canopies are detected through the Local Maximum Method (LMM) and median filter processes.

### 4.4.4 Implementation of Local Maximum Method for detection of tree canopies

In order to determine the required parameter, the local maximums in the 3D point cloud must be calculated with some hypothetical constraints such as using $3 \times 3$ Median Filter, setting tree heights, width and peak to peak distance greater or equal to 1.6 m, 2 m, and 2 m, respectively. This will maximize the accuracy of the model by avoiding many deceptive local maxima. Local maxima are calculated using the following Eq. (16).

$$if\ f(x,y) > f(x_{nn}, y_{nn}) \tag{16}$$

Where: $f(x,y)$ is a pixel of an image, $f(x_{nn}, y_{nn})$ are the neighborhood pixels of the f(x,y). By completing the training of the ANN, $LOS(SSPL_{tree}$ formula is obtained for LOS direction:

$$LOS(SSPL_{tree}[dB]) = FSPL[dB] + \sum_{j=0}^{n_{tree}} \left( \vec{P}L_{tree_j}(h_j, w_j) \right) \tag{17}$$

Where: $\vec{P}L_{tree_j}$: Predicted tree path loss, $h_j$: Height of the detected tree canopy, $w_j$: Width of the detected tree canopy.

#### *4.4.5 Model Validation with MAPE*

Mean Absolute Percent Error (MAPE) is a statistical method to measure the prediction accuracy of models. MAPE determines the differences between real and theoretical values. Later it divides this difference by the real values. Next, the absolute values of the results are averaged and represented as a percentage. MAPE is determined by Eq. (18) [28].

$$LOS(MAPE[\%]) = \frac{100\%}{n} \sum_{i=1}^{n} \left| \frac{Models - PL_{real}}{PL_{real}} \right| \tag{18}$$

## 5. Analysis and results of a case study

This study is based on a comparison of four models, including $PL_{tree}$. The 3D environment obtained by airborne LiDAR belongs to the Florida Institute of Technology neighborhood. Since the data has some faulty values, the natural neighbor interpolation is performed on the raw 3D point cloud. To create a colorful 3D image, the geo-referenced 2D satellite image is imported via Google API and merged with the 3D point cloud. Later, the tree canopies are located by the local maxima method on a 2D image. This 2D classified image helped us extract and locate trees on the 3D point cloud. The process of extracting tree canopies from the 3D point cloud is shown in **Figure 12**.

While the detected trees are marked with red ×, the labeled transmitter and receiver belonging to MCS are marked with yellow + signs. Using the LMM technique and limitations, the height and width of five trees were plotted. The proposed ANN algorithm will be used to estimate the required $PL_{tree}$ value for each tree canopy detected in the LOS direction, where the characteristics of the environment such as height and width are obtained. The properties of five trees, such as distance, width, height, and $PL_{tree}$ are shown in **Table 2**. It is seen from **Table 2**, there is a correlation between tree size and estimated $PL_{tree}$. This is because the complex structure of the tree canopies causes more reflection and absorption in proportion to the tree's size. The detected features from the environment are used as input data for the ANN model, and compared with the other empirical models with the same parameters. The results are listed in **Table 3**. It is seen that the energy demand of LNPL and Cost231 models increased with the distance exponentially, but they could not provide the required signal power considering the real values. This is because traditional propagation models assume that terrains have the same characteristics when it comes to terrain pattern. For this reason, they deviate significantly in terms

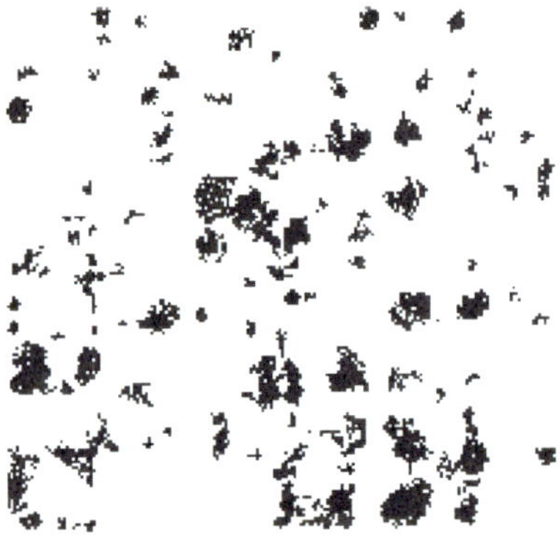

(a) Classified Image

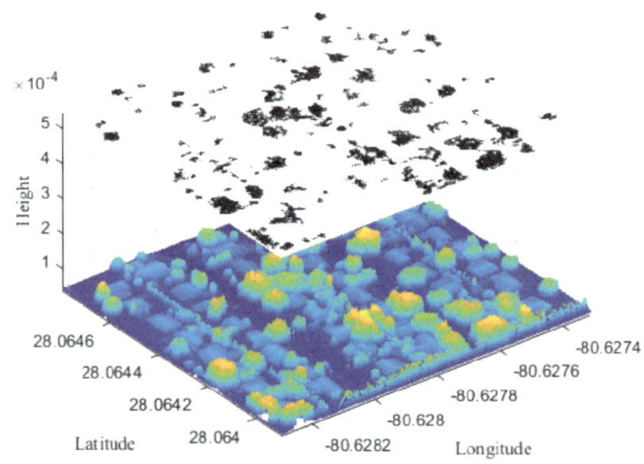


(b) Filtering 3D image using 2D image.

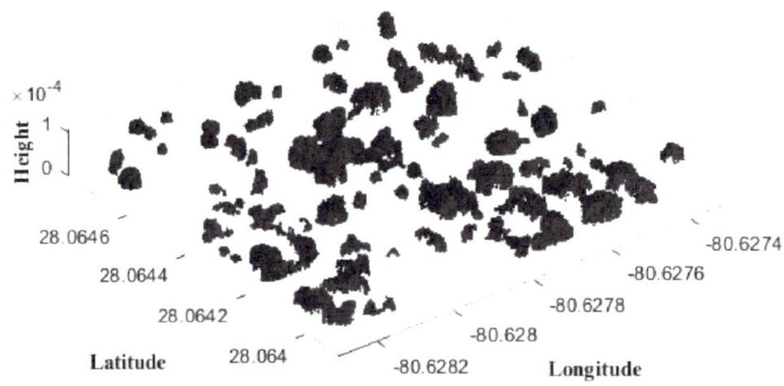


(c) Filtered tree canopies

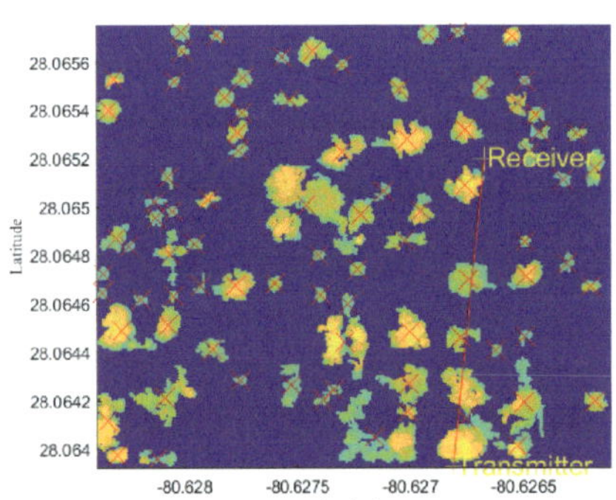


(d) The heat map of 3D extracted tree canopies and Tx and Rx locations.

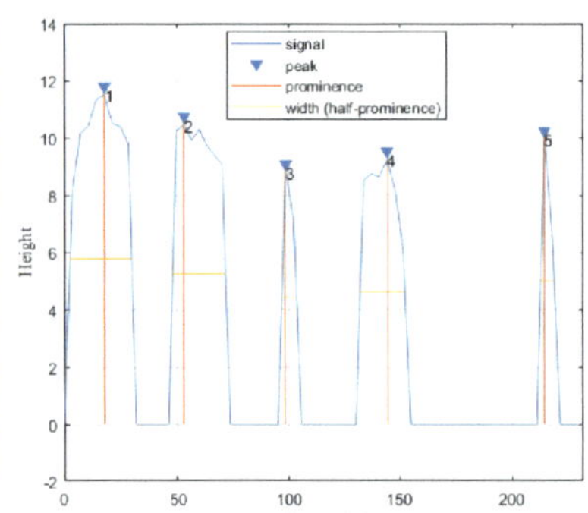


(e) The detected tree canopies and their locations.

**Figure 12.**
*3D Filtering process of tree canopies.*

of SSPL and RSL estimations. The performance of $SSPL_{tree}$ is also shown in **Figure 13**. According to **Figure 13**, All models have a strong relationship in terms of distance and SSPL. However, unlike other models, the tree SPPL is discerning itself by showing a peak anytime propagation encounter a tree in the LOS direction. To see whether the model is performed well or not, RSL measurement should be compared with real values. In this case study, RSL values are taken through transmitter and receiver facing each other with an operating frequency of 2110 MHz. The devices have effective radiated power of -1 dB. All empirical models and SSPL tree predicts the RSL results operating with these parameters. The results are

| Distance($m$) | $n_{tree}$ | Height($m$) | Width($m$) | $PL_{tree}$[dB] |
|---|---|---|---|---|
| 17 | 1 | 11.4 | 13.05 | 3 dB |
| 52 | 2 | 10.45 | 12.15 | 3 dB |
| 98 | 3 | 8.8 | 3.31 | 1 dB |
| 144 | 4 | 9.2 | 10.5 | 3 dB |
| 214 | 5 | 9.9 | 3.15 | 1 dB |

**Table 2.**
*Detected features and obtained* $PL_{tree}$.

| Distance($m$) | $n_{tree}$ | $SSPL_{tree}$[dB] | FSPL[dB] | LNPL[dB] | Cost231[dB] |
|---|---|---|---|---|---|
| 17 | 1 | 49.4 | 46.41 | 48.14 | 30.04 |
| 52 | 2 | 60.93 | 54.93 | 58.26 | 45.05 |
| 98 | 3 | 67.1 | 60.1 | 64.4 | 54.14 |
| 144 | 4 | 73.3 | 63.32 | 68.22 | 59.81 |
| 214 | 5 | 77.7 | 66.7 | 72.24 | 65.72 |

**Table 3.**
*Comparison of Tree Canopy Path loss with empirical models.*

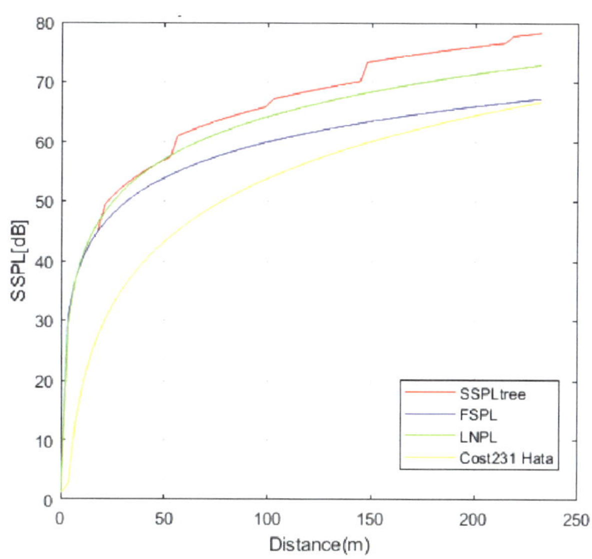


**Figure 13.**
*Comparison of 4 Models with respect to distance.*

presented in **Figure 14**. The RSL values are significant in terms of maintaining the communication between transmitter and receiver. That is why it is essential to keep the predicted values as close as the real values. In **Figure 14**, it is indicated that the red line, which is tree RSL ($RSL_{tree}$), has a similar track with the black line, which is measurement RSL($mRSL$). This proves that the ANN aided model has over-performed among all empirical models. MAPE results also validate these results. From **Table 4**, $RSL_{tree}$ have a minimum deviation of 4.26% in terms of MAPE percentage while others result stayed in between 6.29% to 16.9%.

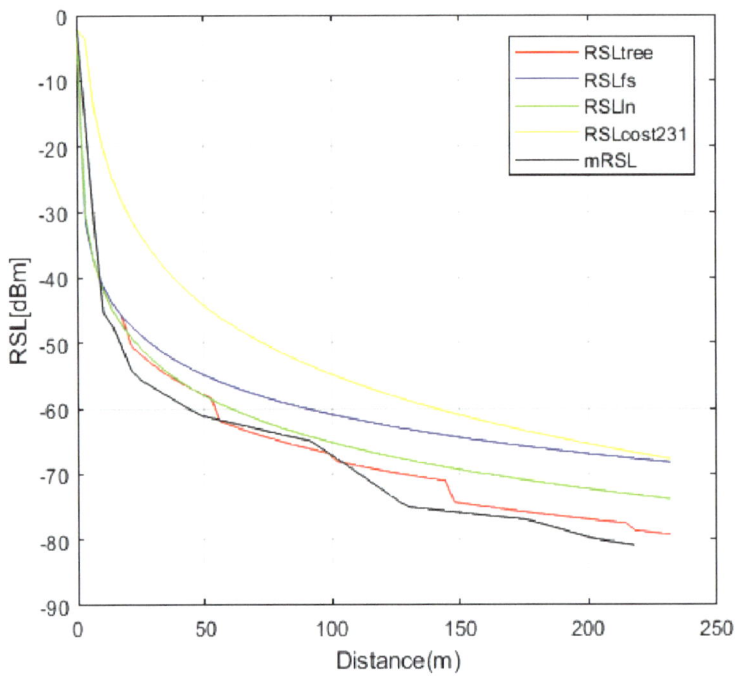


**Figure 14.**
*RSL Comparison of Models and Measured values with respect to distance.*

| | $n_{tree}$ | $RSL_{tree}$ | $RSL_{fs}$ | RSLln | RSLCost231 |
|---|---|---|---|---|---|
| MAPE% | 5 | 4.26 | 10.16 | 6.29 | 16.9 |

**Table 4.**
*The MAPE performance of the models.*

AI aided RSL tree model with the a deviation of 4.26% has a significant improvement compared to other empirical models since all the micro-variations contribute to the estimates. In addition, unpredictability of tree variations [29] was overcome using artificial intelligence. The proposed model result was also outperformed compare to A. Alsayyari et al. case study with MAPE results of % 34.37 and %19.80 [30].

## 6. Conclusions

This chapter demonstrated the fusion of state of art technologies that can potentially contribute to developing an intelligent environment for smarter cities. Sensor fusion, UAV, satellite image, and image classification have integrated for the purpose of creating a 3D virtual environment for a realistic data platform. The obtained information is crucial in terms of the evaluation of the planned projects for futuristic cities. In our case, we have assessed the effect of trees upon smart deployment of MCS using 3D point cloud, which is basically the 3D virtual presentation of the city, to maintain connectivity and efficiency. Since tree canopies are considered irregular terrains and their complex structure highly affect the efficiency of SSPL due to multi-path reflection, we extracted tree canopies by using 2D color classified satellite image as a filter. By means of extracted 3D point cloud information of tree canopies, the features (height and width) are obtained for the purpose of training the ANN. The data and results that are taken from Egi et al.'s case study are a good demonstration of the model application. The analysis shows that the created smart model can significantly affect the MCS' propagation since it adds all micro-variations and utilizes tree features for adjustment of RSL. The MAPE results for all models are obtained as 4.26%, 10.16%, 6.29%, and 16.9% error for $PL_{tree}$, FSPL model, log-normal model, and the Cost 231-Hata model, respectively. It should also be pointed out that the ANN model did not consider the effect of the buildings. This effect may be added to the model in future applications. The primary contribution of this chapter is to create a colorful 3D virtual environment and make more precise feature extraction possible. This technique may also shape the future of smart cities by using digitized information for city planning, communication planning, and infrastructure planning. It should be noted that the proposed model is only applicable to outdoor applications since the 3D virtual environment only provides outdoor information. This limitation can be also removed if the LiDAR scanning is performed indoors and combined with the outdoor data.

## Author details

Yunus Egi[1,2*†] and Engin Eyceyurt[2†]

1 Sirnak University, Sirnak, Turkey

2 Florida Institute of Technology, Melbourne, FL, USA

*Address all correspondence to: yunusegi@sirnak.edu.tr

† These authors contributed equally.

## References

[1] Shirowzhan S, Tan W, Sepasgozar SME. Digital Twin and CyberGIS for Improving Connectivity and Measuring the Impact of Infrastructure Construction Planning in Smart Cities. ISPRS Int. J. Geo-Inf. 2020; **9**:240. DOI: https://doi.org/10.3390/ijgi9040240

[2] Stoll C. Michael Arthur Mehling, COVID-19: Clinching the Climate Opportunity. One Earth. 2020;**3**(4): 400-404, ISSN 2590-3322. DOI: https://doi.org/10.1016/j.oneear.2020.09.003

[3] Qi Q, Tao F, Hu T, Anwer N, Liu A. Yongli Wei. A.Y.C. Nee, Enabling technologies and tools for digital twin, Journal of Manufacturing Systems: Lihui Wang; 2019 ISSN 0278-6125

[4] Shanker Y, Lobiyal DK. Role of earth bulge on the prediction of radio propagation path loss over irregular terrain. International Journal of Wireless and Mobile Computing. 2020; **18**(4):352-360

[5] Carpenter, John, et al. Investigations and analyses of LTE network cell tower deployments and impact on path loss calculations. MITRE CORP BEDFORD MA BEDFORD United States, 2019.

[6] Mir Ghoraishi, Jun-ichi Takada, Tetsuro Imai,"Radio Wave Propagation Through Vegetation" 17 01 2019. [online].Available:http://cdn.intechopen.com/pdfs/42732/InTech-Radiowave propagationthroughvegetation.pdf.

[7] Ribero, MÃ³nica, et al. "Deep learning propagation models over irregular terrain." ICASSP 2019–2019 IEEE International Conference on Acoustics, Speech and Signal Processing (ICASSP). IEEE, 2019.

[8] Getis A. Introduction to Geography: Earth sciences. CTI Reviews: Physical geography; 2016

[9] Cramer M et al. ULTRA-HIGH PRECISION UAV-BASED LIDAR AND DENSE IMAGE MATCHING. Remote Sensing Spatial Information Sciences: International Archives of the Photogrammetry; 2018

[10] Rodrigo Fernandes de Mello; Moacir Antonelli Ponti, Machine Learning"A practical approach on statistical Learning Theory", Sao Paulo: Springer, 2018.

[11] Sherazi, Hafiz Husnain Raza, Luigi Alfredo Grieco, and Gennaro Boggia. "A comprehensive review on energy harvesting MAC protocols in WSNs: Challenges and tradeoffs." Ad Hoc Networks 71 (2018): 117–134.

[12] Juheon, et al. Individual tree species classification from airborne multisensor imagery using robust PCA. IEEE Journal of Selected Topics in Applied Earth Observations and Remote Sensing. 2016;**9**(6):2554-2567

[13] Torabzadeh, Hossein, et al. "Tree species classification in a temperate mixed forest using a combination of imaging spectroscopy and airborne laser scanning." Agricultural and Forest Meteorology 279 (2019): 107744.

[14] Hartling, Sean, et al. "Urban tree species classification using a WorldView-2/3 and LiDAR data fusion approach and deep learning." Sensors 19.6 (2019): 1284. org/citation.cfm?doid=1711113.1711127. [Accessed 6 1 2018].

[15] Thrane J, Zibar D, Christiansen HL. Model-aided deep learning method for path loss prediction in mobile communication systems at 2.6 GHz. Ieee Access. 2020;**8**:7925-7936

[16] Ahmadien O et al. Predicting Path Loss Distribution of an Area from Satellite Images Using Deep Learning. IEEE Access. 2020;**8**:64982-64991

[17] Gracchi, Teresa, et al. "Optimizing Wireless Sensor Network Installations by Visibility Analysis on 3D Point Clouds." ISPRS International Journal of Geo-Information 8.10 (2019): 460.

[18] Egi Y, Otero CE. Machine-Learning and 3D Point-Cloud Based Signal Power Path Loss Model for the Deployment of Wireless Communication Systems. IEEE Access. 2019;7:42507-42517

[19] F. I. University, "International Hurricane Research Center," 12 12 2018. [Online]. Available:http://dpanther2.ad.fiu.edu /Lidar/lidarNew.php.

[20] Liu P et al. Fusion of color histogram and LBP-based features for texture image retrieval and classification. Information Sciences. 2017;**390**:95-111

[21] MATLAB, "Segmentation methods in image processing and analysis," MATLAB, 18 June 2018. [Online]. Available:https://www.mathworks.com/

[22] Breinbjerg, Olav, and Kyriakos Kaslis. "On the accuracy of Friis' transmission formula at short range." 2017 XXXIInd General Assembly and Scientific Symposium of the International Union of Radio Science (URSI GASS). IEEE, 2017.

[23] MacCartney, George R., and Theodore S. Rappaport. "Study on 3GPP rural macrocell path loss models for millimeter wave wireless communications." 2017 IEEE International Conference on Communications (ICC). IEEE, 2017.

[24] Silvia Demetri ; Gian Pietro Picco ; Lorenzo Bruzzone, "Estimating Low-power Radio Signal Attenuation in Forests: A LiDAR-based Approach," in Distributed Computing in Sensor Systems (DCOSS), 2015 International Conference on, Cambridge, 2015.

[25] Braga, Allan Dos S., et al. Radio Propagation Models Based on Machine Learning Using Geometric Parameters for a Mixed City-River Path. IEEE Access. 2020;**8**:146395-146407

[26] de Jong YLC, Herben MHAJ. A tree-scattering model for improved propagation prediction in urban microcells. in IEEE Transactions on Vehicular Technology. March 2004; **53**(2):503-513. DOI: 10.1109/TVT.2004.823493

[27] A. Ng, "Machine Learning lecture 4," 14 01 2019. [Online]. Available: https://www.coursera.org/learn/machine-learning.

[28] Olasupo TO, Otero CE, Olasupo KO, Kostanic I. Empirical Path Loss Models for Wireless Sensor Network Deployments in Short and Tall Natural Grass Environments. in IEEE Transactions on Antennas and Propagation. Sept. 2016;**64**(9):4012-4021

[29] de Jong YLC, Herben MHAJ. A tree-scattering model for improved propagation prediction in urban microcells. in IEEE Transactions on Vehicular Technology. March 2004; **53**(2):503-513. DOI: 10.1109/TVT.2004.823493

[30] A. Alsayyari, I. Kostanic, C. E. Otero and A. Aldosary, "An empirical path loss model for wireless sensor network deployment in a dense tree environment," 2017 IEEE Sensors Applications Symposium (SAS), Glassboro, NJ, 2017, pp. 1–6, doi: 10.1109/SAS.2017.7894099.

Chapter 4

# Digital Twin of the Mining Shaft and Hoisting System as an Opportunity to Improve the Management Processes of Shaft Infrastructure Diagnostics and Monitoring

*Piotr Kalinowski, Oskar Długosz and Paweł Kamiński*

## Abstract

The following chapter presents a concept of a virtual model of a mine shaft equipped with a hoisting system for the purpose of improving the processes of diagnostics management of shaft infrastructure and its monitoring. The chapter presents a proposal of improvement of broadly known processes such as: diagnostics and monitoring of shaft infrastructure using digital models of 3D structures, the BIM and Digital Twin idea. Implementation of such systems in the operating mine working was presented together with expected results of monitoring. As the presented solution is currently only a concept, development of such system in real application is necessary to asses real benefits of application of Digital Twin system.

**Keywords:** mine shaft, data visualization, digital twin, digital model, mine shaft hoisting system, diagnostics, monitoring

## 1. Introduction

Mine shafts along with hoisting systems are ones of the most important parts of underground mines' technological chain. Their correct work is crucial for mine economy and safety of miners and all of the mine's infrastructure. Mine shafts, especially those equipped with hoists, are necessary for the transport of materials, staff, and excavated material. Mine shafts are the only way of rescue from the mine workings. They are also needed for a proper work of mine ventilation system [1].

Mine shafts infrastructure is under constant influence of destructive forces, both geological and anthropogenic origin, which are caused by number of factors, such as local deposit's tectonics (geological structure), rock mass movements caused by pressure effecting from deposit exploitation, atmospheric conditions, groundwater acting on the shaft lining etc. Law and safety regulations ensure that condition of shaft lining and infrastructure is controlled to provide proper levels of safety and efficiency of the shaft. Reliable periodic inventory and tests are made to prevent shaft lining and equipment from destruction. To ensure shaft's safety, despite conducted tests, competent analysis of collected data and quick decision making are needed. It is sometimes necessary to take actions, such as cessation of shaft's operation, repair of shaft equipment elements, which are costly, but necessary to prevent further shaft destruction and to ensure safety of the mine, environment and, most important of all, people's lives and health [2].

Incorrect or unreliable monitoring of shaft infrastructure and data analysis or negligence in those processes can lead to tragic consequences. Proof of how important is proper mine shaft control and monitoring, is an incident, which took place on September 4th 2008 in "Szczygłowice" colliery, part of "Kompania Węglowa S.A.", located in Knurów, Silesian voivodeship, Poland. As the result of this incident, shaft top building of shaft V, main ventilators station, former hoist building, head frame and elements of electricity infrastructure were destroyed [3, 4].

Shaft lining's instability caused sinkhole with radius about 30 metres. Around the sinkhole a danger zone with radius about 100 metres were created. The incident caused abnormalities of functioning of ventilation system both "Szczygłowice" colliery and adjacent "Knurów" mine. In the effect of this, concentration of gases in workings exceeded limits, which was a reason of evacuation of 433 employees of "Szczygłowice" and 92 workers of "Knurów" mine (509 people total). There were no fatalities [5].

Photos below (**Figure 1**) presents sinkhole and remains of the shaft V.

Despite of the rapid character of buildings collapse and sinkhole propagation, the incident was foreseeable. Tests conducted in the shaft V remains showed faults of the shaft lining, which led to the incident. However, incidents which are impossible or almost impossible to foreseen also happen in mines. For example, in R-II shaft of "Rudna" copper mine, part of KGHM Polska Miedź S.A., shaft of the hoisting machine was partially broken in 2011. It resulted in temporal suspension of the mining shaft operation, which brought huge financial loss for the whole company, because hoisting system operating in R-II shaft was one of the most important elements of technological chain not only of the "Rudna" main, but also for the whole KGHM [7].

Photos below (**Figure 2**) presents fracture of the hoist's shaft.

**Figure 1.**
*Sinkhole and remains of shaft V, Szczygłowice coaliery [6].*

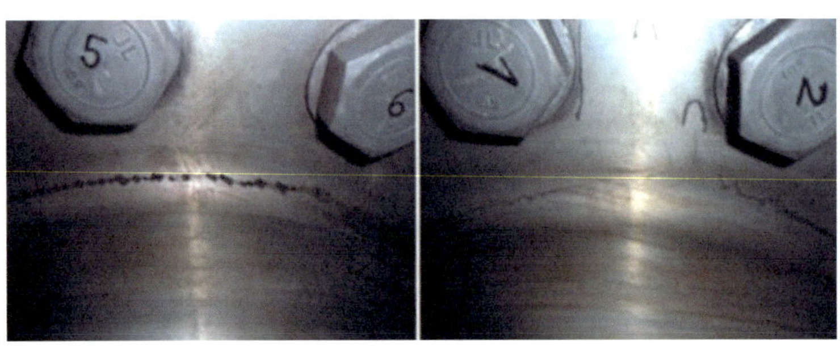

**Figure 2.**
*Fracture of hoist's shaft [8].*

Many different incidents happened in the long history of mining One of their numerous reasons is lack of proper mine shafts control, caused either because of the lack of possibilities of conducting one or negligence in monitoring.

## 2. Control and revision of shaft infrastructure

As it was said before, technical condition of mine shafts' lining and equipment is a foundation for safe underground mine operation. **Figure 3** presents influence of the main workings technical condition on safety of the whole underground mine and its vicinity.

As it was indicated in the introduction, application of Digital Twin system will improve the safety of the whole underground mine. Mine shafts are crucial for maintaining proper mine ventilation and providing transport of people, materials and excavated material. In case of danger, they are the way of staff evacuation, which is essential for people's safety.

It is obvious that condition of particular shaft elements, like hoisting machine, shaft lining or shaft members is reflected in the safety of mine shaft operation. Inappropriate operation of some of these elements can lead to a stoppage of shaft operation, which affects mine's economic performance. In extreme situations, such as those presented in introduction, economic performance of the mine is in a serious danger, because of high cost of claiming the settlement. What is even more

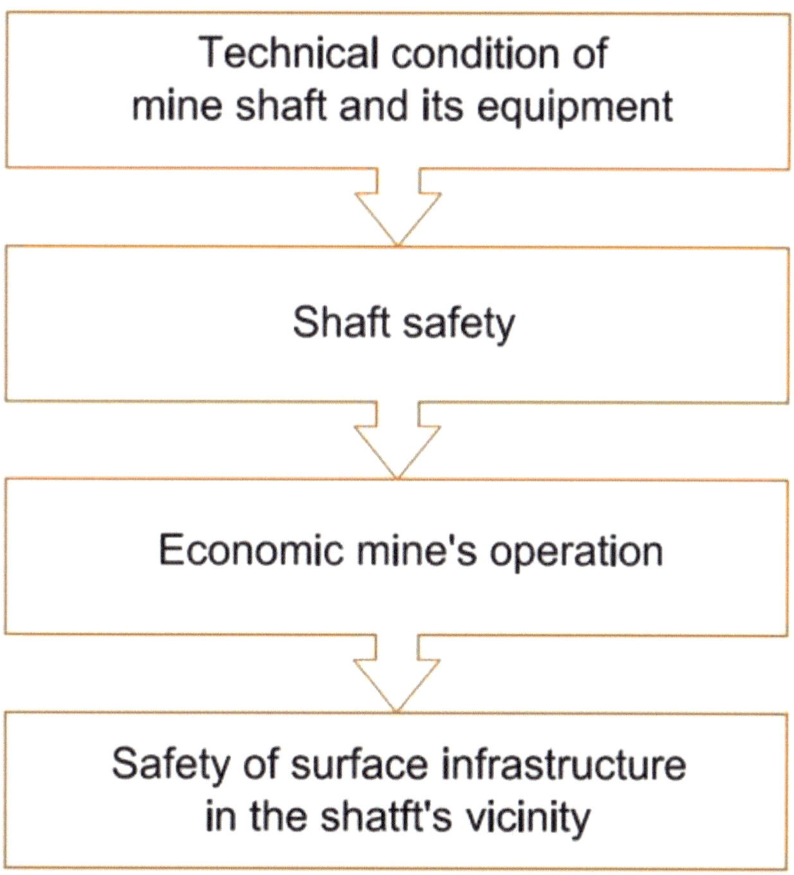


**Figure 3.**
*Influence of technical condition of shaft infrastructure on underground mine's safety.*

important, such situations can cause serious threat for staff of the mine, as well as for bystanders, as mine shafts are sometimes located in the urbanized area.

Instability of shaft's lining can be caused by:

- static load and change in conditions of cooperation of lining and rock mass,
- groundwater flow through the lining,
- deformational load of rock mass,
- dynamic rock tremors,
- utility factors and technological defects,
- aging of lining's material.

Shaft lining damage can lead to serious failures or even disastrous consequences, such as those described in the introduction.

Effects of the shafts' failures can be as follows:

- hazard for employees' lives or health,
- malfunction of ventilation system,
- water, methane or fire hazard,
- shaft failure causing impossibility of its operation,
- damage of buildings located on the surface,
- necessity of temporary suspension of mine operation [1].

According to Polish law regulations mine shafts and their hoisting systems are considered main (literally called "basic") elements of underground mine. Hoisting system with its whole equipment has to be maintained and controlled in a very strict manner, presented in regulations. Use of malfunctioning or broken hoisting system is strictly forbidden by the law [9].

Particular regulations require specific periods of time between specialized inventories of numerous shaft elements by different responsible people. These people are in particular different mine supervisors as well as appraisers.

### 2.1 Disadvantages of currently used systems

Mine shaft is an underground mine's bottleneck, which efficiency is in a relationship with mine's economic performance. Thus, it is desired to take all of the actions to prevent interference of the shaft elements control with its regular operation, as well as its temporal suspensions caused by potential failures or not scheduled maintenance works.

Methods of shaft monitoring are also considered not enough effective nor precise. To solve this problem, idea was presented to, utilizing modern technologies, visualize mine shaft data using specialized measuring methods and software. This way of data collecting, presenting and analysis can have positive impact on mine shaft levels of safety and effectiveness, by reducing time of non-operative work time of the hoisting system [10, 11].

## 3. Data visualization – the idea of digital twin of the mine shaft

During analysis of current state of art of mining, automatic and information technology, idea of mine shaft's digital twin construction was born. Mine shaft equipment, with numerous monitoring devices, which data can be collected, analyzed and processed in real life, can effect in more reliable and accurate forecasts of failures or stoppages of shaft operation. The key factor of this idea's success is integrity of applied solution. To provide full and complex data, it is needed to cover different elements of shaft elements monitoring. The most important of them is visual examination (using video cameras). The other aspects to analyze are power consumption, temperature of particular hoisting system elements, season, head frame deflection, etc. [12, 13].

The main problem to be solved is to state relationships between many, seemingly unrelated, factors. Future goal is to make the digital twin independent structure with decision making mechanism, to decide about the parameters of needed maintenance works or necessary stoppages. To achieve this goal, it is essential to spend time to "teach" the machine how to make appropriate and safe decisions, compliant with law regulations and experience of mine management and engineers. Having knowledge and experience, the machine should analyze collected data and their influence on other factors, as well as compare their quantities with limit values [14].

## 4. Devices for data visualization

### 4.1 Digital engineering solutions

Technologies developed by DES are engineering tools utilizing artificial intelligence for ACE market (architecture, construction and engineering) and asset management. DES' solutions are also present on NDT market (non-destructive testing), as well as BIM and EAM.

Software developed by *Digital Engineering Solution* utilizes pictures and files taken by drone, phone or camera. Advanced algorithms provide image processing, which allows to observe changes on monitored objects, such as size and location of fracture (**Figure 4**), elements displacement, deflection etc. Collected data can be stored and evaluated to assess risk and reliability of analyzed solutions [15].

Characteristic features of DES solution are:

- utilizing smartphones as measurement devices,
- processing of photos and videos for infrastructure monitoring and damage detection (using artificial intelligence),

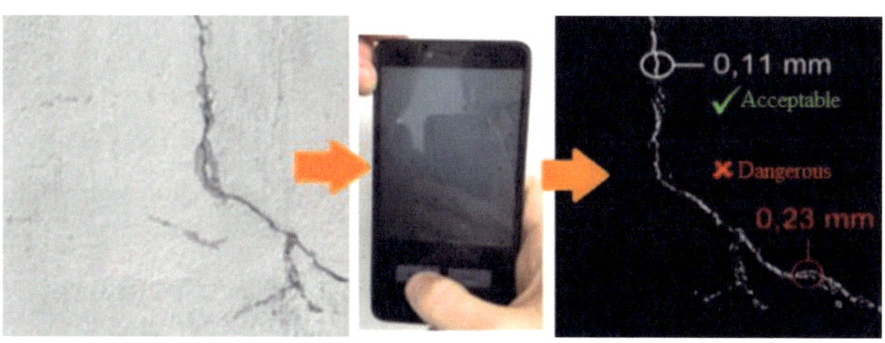


**Figure 4.**
*Example of DES measurement application [15].*

- drafting 3D BIM models from photos (taken with smartphone or any other device, such as intrinsically-safe camera ATUT), as well as their updating with current test data (to assist in decision making process).

Use of DES platform allows to decrease time needed for monitoring process initiation from days to minutes. DES innovative solution is an answer for global need for low-cost monitoring systems. In terms of pandemic crisis and quick technological advance it is a real issue [15].

### 4.2 ATUT, intrinsically-safe system

Specific environment of the monitored object force usage of devices with particular features and complying certain safety standards. Such solutions, which meet mining industry safety standards are produced by Polish company PPHU ATUT sp. z o.o., so they can be used to extend potential of digital twin.

AT VIDEO system is used to obtain video footage from hard-to-reach places, as well as from areas which are particularly dangerous, in which attendance of people should be restricted. ATUT system support multiple video cameras at the same time and data transmission of digital image in mine workings and on the surface, using fiber-optic Ethernet network. Such solution enables reduction of fibers number.

ATUT, intrinsically-safe video system consists of:

- AT-NODE/G – node of redundant fiber-optic backbone,
- PZW-1/ATViso-2 – surface visualization unit using ATVisio-2 software,
- ISE-1 – intrinsically-safe Ethernet network switch,
- KM-2 – media converter,
- DZW-1 – underground fireproof computer,
- IKA-1 – intrinsically-safe video camera.

IKA-1 video camera is a standard system element. This camera is equipped with automatic aperture control, regulating light intensity. In case of extremely-low illumination the camera sets itself in the black and white mode. Recorded image is transferred by fiber-optic network or coaxial cable with tele-technical twisted pair. If utilization of fiber-optic installation is impossible, fiber-optic video converter might be used [16].

Diagram of the system's idea is shown in **Figure 5**.

There are also other solutions utilizing data visualization available on the market, such as:

- ATUT-RFID – system that allows assessment of type and mass of materials transported with the hoist influence on correctness of hoisting system work,
- AT-Location – system used to localize mine workers both in underground workings and on the surface (as well as during transportation in man shaft). Data is collected in near-real-time and saved [17].

*DOI: http://dx.doi.org/10.5772/intechopen.96193*

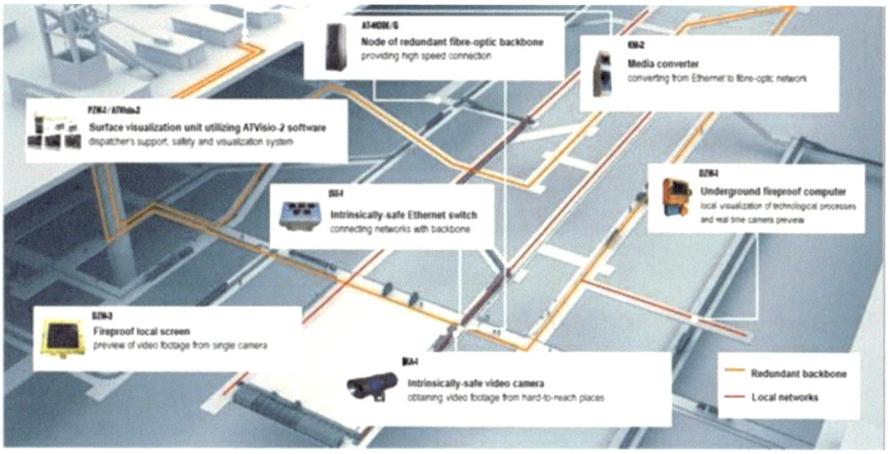


**Figure 5.**
*Idea of AT-VIDEO system application in an underground mine [16].*

### 4.3 3D scanning

In the last few years one can see rapid development of laser scanning technology. Numerous researches and tests are carried out on its use for inventories. *Skala 3D* company introduced mobile system for automatic mine shafts geometry measurement, which provides full and precisely mapped model of the object. It is based on data collected by laser scanners and inertial unit. Use of GPS is impossible for measurements carried out with the system, because tests are conducted underground. Thus, trajectory of scanners' motion is determined by geometrical data, accelerometers and gyroscopes of inertia unit. System is also equipped with set of vibroisolators, to prevent influence of platform's vibrations during its movement in the shaft [18, 19].

The whole system presented above is a fully calibrated measurement unit, which can provide spatial data from measurements of analyzed shaft in relatively short time. Accuracy of measurement is about 2–3 mm in one measurement plane. The system is considered accurate enough for in situ tests in mine shafts [19].

### 4.4 GPS measurements

In specialized literature one can find multiple examples of use of GPS measurements undertaken for civil engineering and also for mining engineering. One of them is its use for Polish coal mine LW Bogdanka, where in year 2012, system of head frames tops displacements monitoring, applied by Department of Geomechanics, Civil Engineering and Geotechnics of AGH UST in Krakow. Final effects were satisfactory. Broad range of GPS use possibilities as well as positive research experience indicates that this technology might be very useful for similar purposes [20]. **Figure 6** presents head frames of Bogdanka's shafts with GPS antennas mounted on their tops.

### 4.5 Other devices

To provide full control of mine shaft infrastructure numerous other devices can be used. Examples of such devices are thermal imaging cameras installed on different elements of shafts or hoisting systems to monitor temperature of breaks,

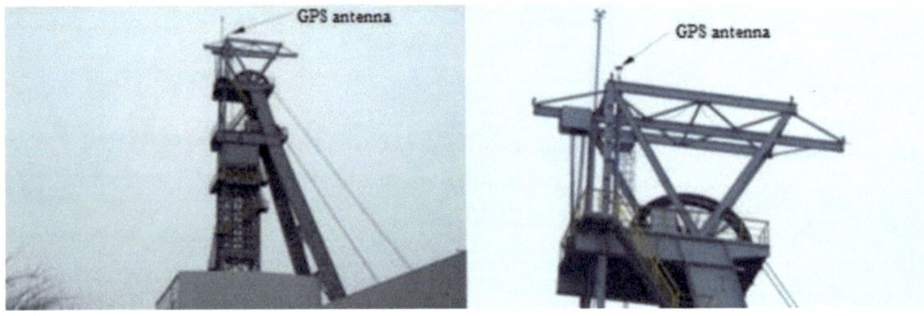


**Figure 6.**
*Head frame of S1.2 shaft, LW Bogdanka coaliery.*

hoist shaft, etc. Such applications might be useful for control of hoisting system performance. Thermal cameras systems for mine shaft's applications must be intrinsically safe and have wide measuring range, to provide safe work during all phases of shaft's operation, including its sinking (with utilization of rock mass freezing).

### 4.6 Geotechnical monitoring of rock mass and mine shaft lining

Proper shaft lining monitoring is crucial to provide safe shaft operation. It is required by Polish law on every phase of shaft existence – its sinking, operation and liquidation. Tests cover several parameters of shaft elements and equipment, such as:

- condition of shaft lining, defined by observations, destructive and non-destructive tests,
- shaft's lining integrity,
- lining's stability,
- correctness of shaft members installation,
- shaft elements' state of wear [21].

Shaft lining monitoring can be carried out by:

- *Vibrating Wire Stressmeter GEOKON,*
- *NATM Shotcrete Stress Cell Geokon,*
- *Concrete Stressmeter Geokon,*
- *Smartec SOFO Standard Deformation Sensor,*
- *NSM/ENSM system by Elexon Mining.*

One of the most important factors of proper mine shaft's monitoring is arrangement of measurement devices. Such theoretical devices arrangement is shown in **Figure 7**.

### 4.7 Idea of monitoring system

Complex graph of idea of monitoring system is presented in **Figure 8**.

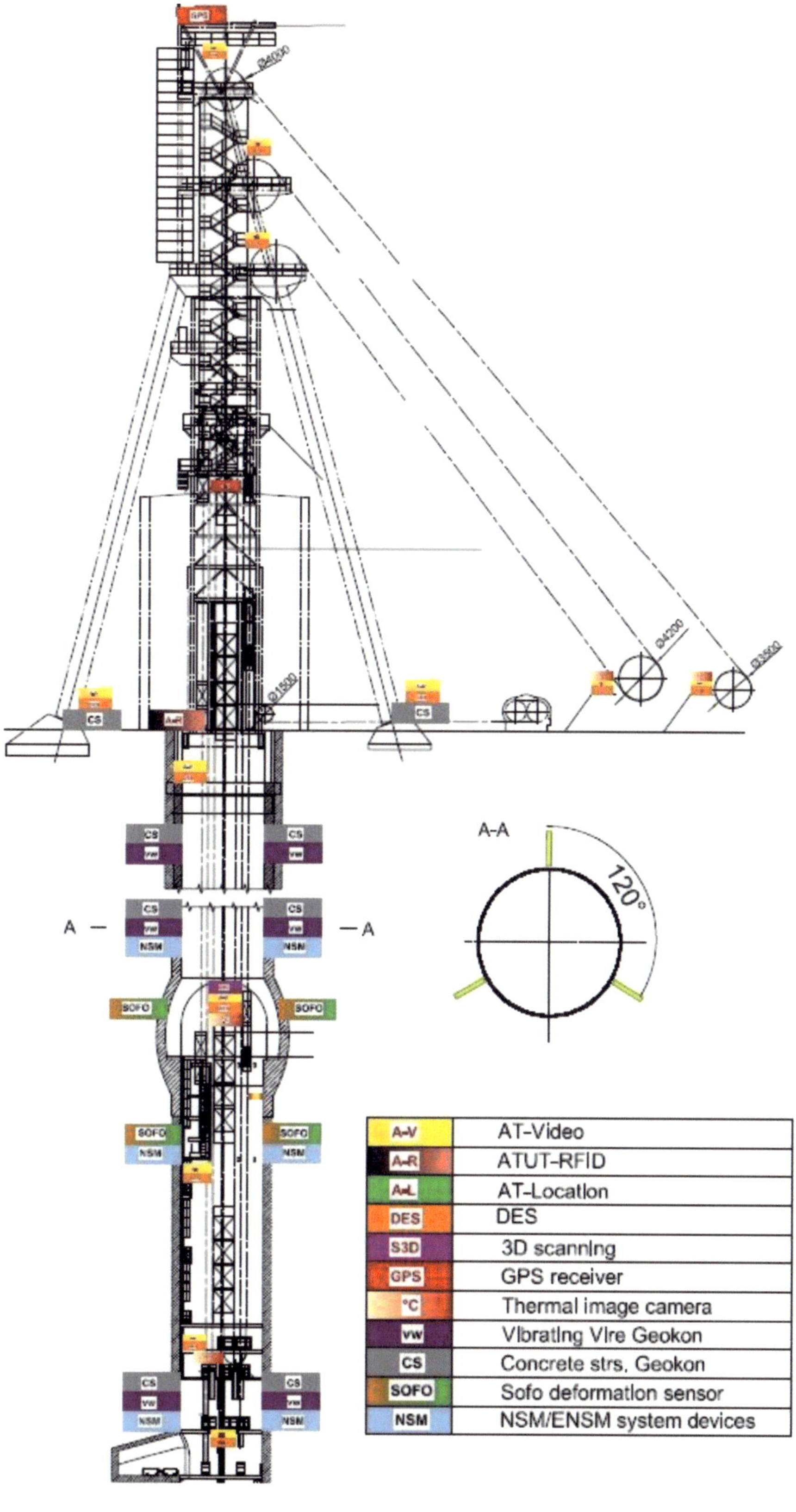


**Figure 7.**
*Example of devices' arrangement for mine shaft monitoring.*

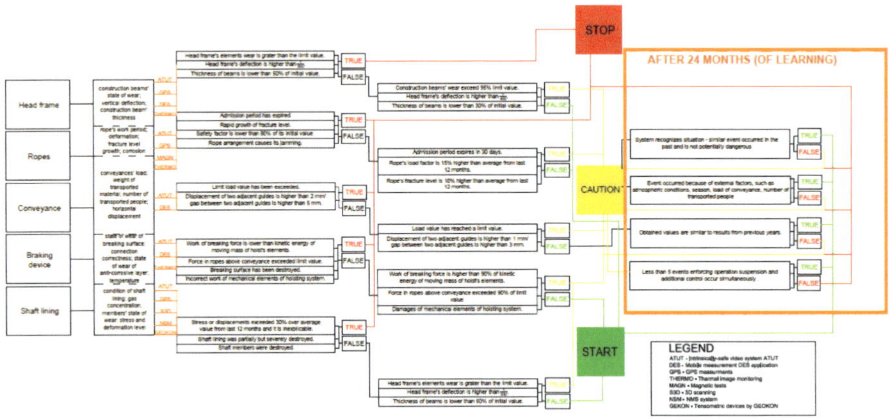


**Figure 8.**
*Idea shaft lining and hoist monitoring system.*

## 4.8 Management of diagnostics

To provide effective operation of a shaft and its infrastructure, proper planning of all suspensions, maintenance works, revisions, controls and inventories is needed. Digital twin of mine shaft might be useful for that purpose. However, to ensure complete diagnostics management it is worth employing models and applications used in civil engineering. Their utilization in mine shaft monitoring should significantly improve safety and economic performance of the mine. Application proposed for diagnostics management of the shaft infrastructure as well as the whole mine are presented below.

*BIM* (Building Information Modeling) – is a process supported by various tools, technologies and contracts involving the generation and management of physical and functional characteristics of places. Parametrical data is gathered to provide information about analyzed infrastructure. BIM is a tool for generating building data, its designing and management during operation. BIM's characteristic feature is easy access to data, so all stakeholders can have access to the same complete information. Parametric data record ensures possibility of computer modeling of building (basing on tables, calculations, data analysis, etc.). Use of BIM goes beyond the planning and design phase of project, extending throughout the building life cycle. The supporting processes of building lifecycle management includes cost management, construction management, project management, facility operation and application in green building. The most popular BIM software are programs Autodesk Revit and Graphisoft Archicad.

BIM models are visually attractive but not automatically updated, so they do not contain current data about state of the infrastructure or its damages. Thus, it is recommended to extend BIM models using other technologies. Digital data of BIM model is often supported using additional applications. Some of these applications are presented below.

## 4.9 Mobile applications in civil engineering

***Doxcel*** utilizes software based on artificial intelligence for image analysis to give real-life information about timeliness of scheduled operations, budget implementation or even quality of work at construction site.

***Building System Planning*** is a solution based on automation of civil engineering project planning. One of its elements is GenMEP, software used for automation of mechanics, electric and hydraulic installation design, in terms of BIM. For example, GenMEP can automatically and autonomously design different installations for building of which its 3D model was previously made. Purpose of use of such software is to prevent colliding several installation elements in one place.

***Autodesk BIM 360 DOSC*** is an application for management of documents on construction site, which is also adapted for teamwork. There are also similar software solutions available on market, e.g., PLANGRID and PROCORE. Usually they provide photo documentation management, but without advanced image analysis. Their most important features are reporting, time and budget management. They also support management of contacts, meetings, deliveries, etc. There are also other applications for individual clients, such as mobiDOM, which consist of schedule, organizer, contact database, etc.).

Digital Engineering Solutions in cooperation with Przedsiębiorstwo Budowy Szybów S.A. (Shaft Sinking Company, part of JSW Group) developed an application

**Figure 9.**
*Model of mine shaft and hoisting system made with DES's application.*

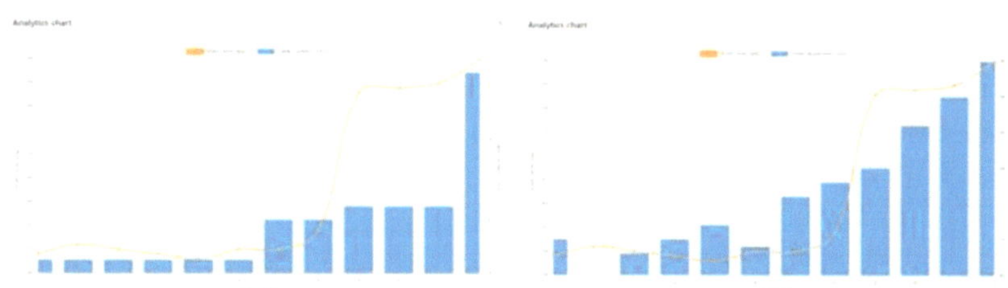

**Figure 10.**
*Graphs of shaft elements' data and relationships between them.*

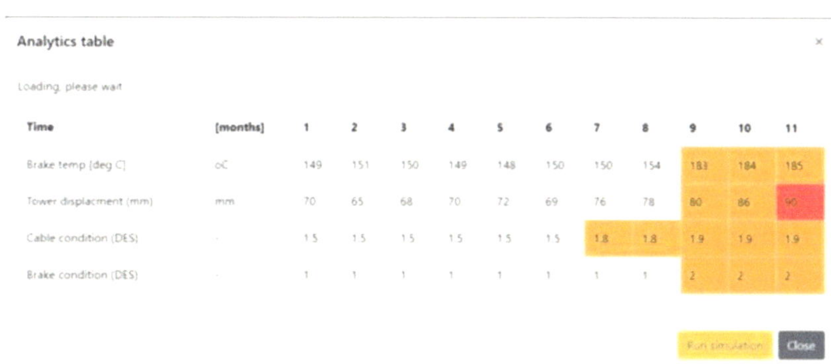

Analytics table

Loading, please wait

| Time | [months] | 1 | 2 | 3 | 4 | 5 | 6 | 7 | 8 | 9 | 10 | 11 |
|---|---|---|---|---|---|---|---|---|---|---|---|---|
| Brake temp [deg C] | oC | 149 | 151 | 150 | 149 | 148 | 150 | 150 | 154 | 183 | 184 | 185 |
| Tower displacment (mm) | mm | 70 | 65 | 68 | 70 | 72 | 69 | 76 | 78 | 80 | 86 | 90 |
| Cable condition (DES) | - | 1.5 | 1.5 | 1.5 | 1.5 | 1.5 | 1.5 | 1.8 | 1.8 | 1.9 | 1.9 | 1.9 |
| Brake condition (DES) | - | 1 | 1 | 1 | 1 | 1 | 1 | 1 | 1 | 2 | 2 | 2 |

Run simulation Close

**Figure 11.**
*Table of shaft element's data.*

prototype for digital twin development. The concept and real-life application possibilities are currently an objective of further development. Picture above (**Figure 9**) presents an example of a model made with this application.

Presented application enables user to present data in graphical form, as well as show relationships between them. **Figure 10** presents graphs and relationships of brake temperature and rope condition (on the left side) and break temperature and tower displacements (on the right-hand side). In the **Figure 11** table made with DES's app and presenting data listed above is shown. Orange and red color indicates dangerous states.

## 5. Discussion

**Table 1** below presents simplified SWOT analysis of proposed solution of Digital Twin. Obvious advantages of such system are improvement of safety in the mine workings in which monitoring is to be used. Improvement of economic performance of the mine is possible only if investment costs and spending on training or hiring qualified staff are not too high. However, it is very hard to assess real saving achieved by application of Digital Twin. It might be real issue, because saving in case of situations similar to those presented in introduction might be huge, but in other cases they can be as well unnoticeable.

Idea of Digital Twinning is a pioneering solution in Polish mining industry. The coal-based industry suffers from underinvestment, which leads to a situation in which the whole industry is out-of-date. As the $CO_2$ emission allowances' prices are constantly raising, it might be hard to introduce such system in a dying industry. A chance for development of Digital Twin system is an industry of raw materials, such as copper etc., because demand for such resource is constantly growing.

| Strengths | Weaknesses |
|---|---|
| • improvement of the safety in the mine shaft and its vicinity,<br>• improvement of mine's economic performance,<br>• increase of the automation level of the mine | • high investment cost,<br>• demand for qualified staff,<br>• problematic installation in operating shaft |
| **Opportunities** | **Threats** |
| • possibility of detailed analysis of gathered data,<br>• possibility of conducting research basing on archived data,<br>• possible development of the system, covering other mine objects or different branches of industry,<br>• potential financial profit | • a conflict with some regulations (necessity of adjustment of law regulations),<br>• necessity of employment or training highly qualified staff |

**Table 1.**
*Simplified SWOT analysis of digital twinning.*

## 6. Conclusion

In the chapter, idea of comprehensive monitoring system of mine shaft infrastructure, as well as tools supporting diagnostics management were introduced. Presented solution utilizes modern technologies, including BIM and Industrial Internet of Things. Main goal of digital twin of mine shaft is reduction of unscheduled suspensions of shaft operations and improvement of its safety, by constant shaft elements monitoring with analysis of gathered data. Mutual influence of different shaft elements and their impact on reliability of its operation can be also done basing on data collected during control process. In long time period it might reduce costs of shaft maintenance.

Theoretical model of mine shaft monitoring was prepared according to existing Polish law. Goal of this venture is to prepare comprehensive visualization of mine shaft and its equipment, to provide virtual analysis of its behavior. Monitoring system should be able to learn how to react for different events occurring in shaft elements, using systems such as ABB Ability™ or similar (SYSTEM PI) and proper monitoring unit. Decisions made by software might be displayed on responsible person's computer screen. Systems such as DES consists of digitization both underground and surface infrastructure. In future it can provide assessment of mining damage, using scanning, modeling, EAM integration and, as an effect, introduction of BIM and EAM mechanisms. Such solutions might help reducing amount of money spent by mining companies to handle mining damage, which is real issue, because in 2018 only JSW Spent 92 million zł (approximately 25 million USD) for this. Presented idea is a only a theoretical solution, so it's real life applications needs further analysis to determine amount of measuring devices, their location and performance in hard conditions of mine shaft. Digital model of mine shaft and hoisting system has a positive impact on economical effectiveness of mine and shaft itself, as well as safety of mine staff and infrastructure. On the side of disadvantages there are high investment cost, long time of system introduction and its "learning" and reliability of applied devices and systems. However, in long time perspective costs presented above are much lower than losses resulting from unscheduled suspensions of shaft operation. Improvement of people's safety is also very important result of digital twin application.

## Author details

Piotr Kalinowski[1], Oskar Długosz[1] and Paweł Kamiński[2]*

1 Przedsiębiorstwo Budowy Szybów S.A., ul. Hutnicza, Tarnowskie Góry, Poland

2 Faculty of Mining and Geoengineering, AGH University of Science and Technology, al. Mickiewicza, Krakow, Poland

*Address all correspondence to: pkamin@agh.edu.pl

## References


[1] Czaja P.; Kamiński P. *Wybrane zagadnienia technik i technologii drążenia szybów*. Kraków: Szkoła Eksploatacji Podziemnej, 2016.

[2] M. Szade M.; Szot A. *Techniczne metody kontroli podstawowych obiektów zakładu górniczego.* Prace Naukowe Gig Górnictwo i Środowisko. 2006, 3.

[3] Wyższy Urząd Górniczy wug.gov.pl. [Online] 09 2008. [cited: 18 05 2020.] http://www.wug.gov.pl/bhp/04_09_2008.

[4] Wyższy Urząd Górniczy. wug.gov.pl. [Online] 09 2008. [cited: 18 05 2020.] http://www.wug.gov.pl/o_nas/wiadomosci_wug/Katastrofa-budowlana-w-KWK-Szczyglowice/idn:137.

[5] Baca-Pogorzelska Karolina. Rzeczpospolita. rp.pl. [Online] 31 03 2009. [cited: 18 05 2020.] https://www.rp.pl/artykul/284531-Zawalenie-szybu-w-Szczyglowicach-to-wina-czlowieka-.html

[6] nettg.pl. [Online] 26 02 2010. [cited: 18 05 2020.] https://nettg.pl/news/14148/sep-2010-dlaczego-runal-szyb-piaty-.

[7] Wyższy Urząd *Górniczy Decyzja Urzędu Górniczego do Badań Kontrolnych Urządzeń Energomechanicznych. L. dz. UGB/0232/0001/11/01520/Sz.,*

[8] Kowal L. Instytut Techniki Górniczej KOMAG. Maszyny Górnicze. 2013, 2

[9] Rozporządzenie Ministra Energii z dnia 23 listopada 2016 r. w sprawie szczegółowych wymagań dotyczących prowadzenia ruchu podziemnych zakładów górniczych

[10] Jabłoński M Jaśkowski W. *Przegląd technik inwentaryzacji rury szybowej. Budownictwo i Architektura*. 2016, 15(3).

[11] Kaleta H *Założenia systemu monitorowania szybów górniczych w świetle wybranych uszkodzeń obudowy szybów*. Systemy Wspomagania W Inżynierii Produkcji. 2017, 6.

[12] Battista N. Cheal, R. Harvey, C *Monitoring the axial displacement of a high-rise building under construction using embedded distributed fibre optic sensors*. Kechavarzi. 2017.

[13] Matthew N. O. Sadiku, Adebowale E. Shadare, Sarhan M. Musa *Information Engineering,* International Journal of Engineering Research, Volume No.6, Issue No.11, pp: 448-449, Oct. 2017

[14] Matthew N.O. Sdiku, Adebowale E. Shadre, Sarhan M. Musa, Cajetan M. Akujuobi *Data Visualization*, International Journal of Engineering Research and Advanced Technology, Volume. 02 Issue.12, December-2016

[15] Bednarski G. *Demonstracja wykorzystania mobilnej aplikacji pomiarowej w Asset Management* 2020

[16] PPHU ATUT Sp. z o.o. *AT - VIDEO Advertising materials of PPHU ATUT Sp. z o.o*. Mysłowice 2019.

[17] PPHU ATUT Sp. z o.o. *SMC-1/KWP-1 Advertising materials of PPHU ATUT Sp. z o.o.* Mysłowice, 2019.

[18] Adamek A.; Skala 3D *Mobilna platforma górnicza (MPG) - nowatorskim rozwiązaniem w polskich kopalniach*. Archiwum Fotogrametrii, Kartografii i Teledetekcji. 2015, 27

[19] Preuss R. *Automatyzacja procesu przetwarzania danych obrazowych*. Archiwum Fotogrametrii, Kartografii i Teledetekcji. 2014, 26.

[20] Tajduś A.; Stewarski E.; Kamiński P. *Monitoring satelitarny GPS mikroprzestrzeni szczytów wież szybowych*

*DOI: http://dx.doi.org/10.5772/intechopen.96193*

*w kopalni LW,,Bogdanka"*. Kraków: Akademia Górniczo-Hutnicza, 2012.

[21] Calikowski B. Bielceka.R. *Odkształcenia, naprężenia, przemieszczenia i temperatury w obudowie szybu zmierzone aparaturą tensometryczną typu, SZAT-1"*. Zjednoczenie Budownictwa Kopalń Rud. 1960.

Section 3

# Machine Learning and Future of Data Science

Chapter 5

# Using Trend Extraction and Spatial Trends to Improve Flood Modeling and Control

*Jacob Hale, Suzanna Long, Vinayaka Gude and Steven Corns*

## Abstract

Effective management of flood events depends on a thorough understanding of regional geospatial characteristics, yet data visualization is rarely effectively integrated into the planning tools used by decision makers. This chapter considers publicly available data sets and data visualization techniques that can be adapted for use by all community planners and decision makers. A long short-term memory (LSTM) network is created to develop a univariate time series value for river stage prediction that improves the temporal resolution and accuracy of forecasts. This prediction is then tied to a corresponding spatial flood inundation profile in a geographic information system (GIS) setting. The intersection of flood profile and affected road segments can be easily visualized and extracted. Traffic decision makers can use these findings to proactively deploy re-routing measures and warnings to motorists to decrease travel-miles and risks such as loss of property or life.

**Keywords:** trend extraction, spatial and temporal trends, images

## 1. Introduction

Floods are the most frequently occurring natural disaster. A flood event occurs when stream flows exceed the natural or artificial confines at any point along a stream [1]. This is often due to heavy rainfall, ocean waves coming on shore, rapid snow melting, or failure of manmade structures such as dams or levees [2]. From 1998–2017, flood events affected more than two billion people globally [3]. Disasters of this frequency and magnitude are typified by extreme costs to governments. In 2019, historic flooding across Missouri, Arkansas, and the Mississippi River basin resulted in an estimated cost of 20 billion dollars [4]. These estimates typically do not reflect indirect costs such as added travel-miles and the subsequent loss of time. Further, floods are among the deadliest natural disasters. From 2010–2020, floods resulted in the fatalities of 1089 people in the United States [5]. A majority of these deaths were comprised of motorists. Therefore, urban planners such as traffic decision makers are tasked with proactively deploying resources that minimize motorist risk exposure. At present, traffic decision makers rely on static flash flood inundation profiles related to discrete rainfall events. These profiles are often created through multiagency cooperation efforts such as [6]. Some studies have begun to generate dynamic flood inundation data visualizations based on these profiles [7].

Additionally, integrated approaches that use machine learning and geographic information systems (GIS) to track changes in critical infrastructure over time are emerging as powerful decision support tools [8]. However, there is limited use of state-of-the-art time series prediction models to generate dynamic data visualizations in a GIS setting for improved flood management. This book chapter explores the integration of publicly available data and machine learning models to address this gap in the literature.

Precise determination of when and where to deploy re-routing measures is a complex task. One approach that improves planning effectiveness is to integrate time series characteristics of river behavior and corresponding spatial flood profile. In this chapter, a univariate time series prediction of river stage is conducted that improves the temporal resolution and accuracy of publicly available forecasts. This prediction is then tied to a corresponding spatial flood inundation profile in a GIS setting. The resulting geospatial deep learning model provides a data visualization tool that traffic decision makers can use to proactively manage road closures in the event that a flood is likely to occur. The first section provides an overview of relevant river behavior that causes flooding. State-of-the-art trend extraction and prediction techniques are then presented and tied to geospatial use cases. The methodology section presents the data used, time series prediction model selected, and geoprocessing procedures required for data visualization using GIS software. Next, an illustrative example is provided for a frequently flooded intersection in Missouri. A discussion section is provided that positions the findings in the context of improving traffic management in the event of a flood. Lastly, a conclusion is given that summarizes the key findings and outlines model limitations and future work.

## 2. A geospatial deep learning approach

Two key characteristics of streams that relate to flood events are stream stage and streamflow. Stream stage refers to height (ft) of the stream and streamflow corresponds to discharge ($ft^3/s$) or alternatively, volumetric flowrate. Typically, governmental organization such as the United States Geological Survey maintain a network of sensors that monitor these characteristics over time for various stream segments. The National Weather Service classifies flood categories into four groups based on stream stage: Action Stage, Flood Stage, Moderate flood Stage, and Major Flood Stage [9]. These values vary for a given segment of stream based on analysis of previous floods, local topography, and underlying geological properties.

Given that stage is monitored over time, the use of time series forecasting methods to predict stage values is appropriate. There are two modeling approaches that are useful in this context: statistical and computational intelligence. Statistical models use historical data to identify underlying patterns to predict future values [10]. Some commonly used techniques for flood forecasting include simple exponential smoothing [11], autoregressive moving average [12], and autoregressive integrated moving average [13]. However, one shortcoming of these approaches is lack of scalability as the quantity and complexity of data increases [14]. An alternative approach that addresses these issues is computational intelligence. A key feature of computational intelligence approaches is the capacity to manage complexity and non-linearity without needing to understand underlying processes [15]. In summary, statistical methods rely on precise underlying relationships and exhibit decreased performance as the number of variables increases whereas computational intelligence approaches identify patterns using large amounts of training data to establish a model capable of accurate predictions [16]. Some commonly used flood forecasting computational intelligence models include support vector machines [17], artificial neural networks [18], and deep learning [19]. Further, they have demonstrated superior performance when compared to conventional statistical modeling approaches for flood prediction studies. LSTM models have explicitly shown promising results in time series contexts. Therefore, LSTM models provide a state-of-the-art trend extraction and prediction technique regarding stream stage values.

Stream stage values are categorized based on resulting flood severity. The physical reality of these categories is the spatial extent of the flooding event often referred to as a flood inundation map [20]. These maps provide decision makers with a useful visual reference to determine what specifically has been affected by a flood event. An area of research, data visualization, and practical application that has not been fully investigated is the integration of computational intelligence stream stage predictions with geospatial flood inundation maps. The methodology provided in the following section addresses this gap.

## 3. Methodology

This section consists of three parts: LSTM prediction of stream stage, data required, and geoprocessing procedures. First, a brief overview of LSTM will be given. This will include explanatory figures and relevant mathematical formulas. Second, data required to conduct the LSTM prediction of stream stage will be procured. Flood inundation imagery and road network data will also be obtained. Lastly, data will be uploaded to a GIS software and processed for end use by traffic decision makers. An illustrative example is presented in the next section.

### 3.1 LSTM prediction of stream stage

Stream stage prediction is a time series forecasting procedure that is dependent on previous data to predict future values. As the quantity and quality of data continues to increase, more powerful computational approaches can be applied to prediction problems. The results of the literature review demonstrated that deep learning approaches, namely LSTM networks, are increasingly being applied to these problems.

Deep learning is an extension of the conventional neural network by adding additional layers and layer types. **Figure 1** provides a visual comparison of the two approaches [21]. The simple neural network (left) consists of a single input layer, hidden layer, and output layer. Alternatively, the deep learning neural network (right) has one input layer followed by three successive hidden layers that ultimately feed into a final output layer. This configuration has generated superior performance in capturing complex relationships.

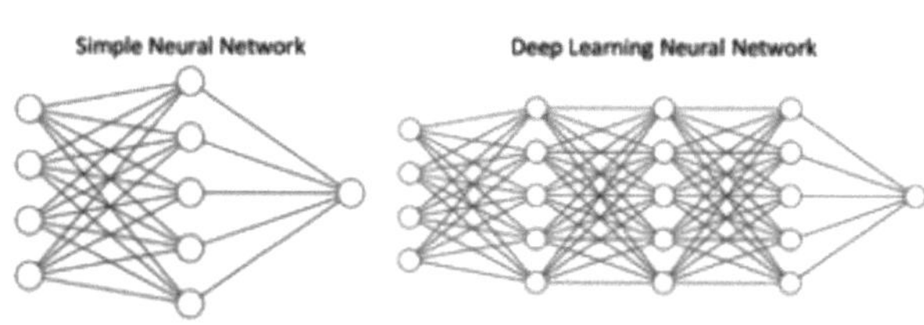


**Figure 1.**
*Simple neural network vs. deep learning neural network.*

However, neither approach retains previous time step information. Recurrent neural networks (RNNs) were introduced to address this limitation. LSTM networks are the deep learning variant of RNNs. All figures and mathematical formulation are borrowed from [15]. The primary benefit of LSTM networks is the capacity to retain longer term information. This is accomplished by removing and adding information determined by a series of 'gates' and vector operations. **Figure 2** provides a visual representation of an LSTM cell. The first gate, illustrated in yellow, generates a value between 0 and 1 using the current input ($x_t$) and output from the previous step ($y_{t-1}$) that determines how much information is passed on (forget gate). A zero corresponds to no information transfer whereas a one represents a complete transfer.

The result of this procedure ( $f_t$ ) is presented mathematically in Eq. (1) as a sigmoid neural network layer where U (weights) and W (recurrent connections) are matrices.

$$f_t = \sigma\left(x_t U^f + y_{t-1} W^f\right) \quad (1)$$

Next, a decision must be made regarding what information needs to be stored. This is accomplished by applying an additional sigmoid layer (red, $i_t$). New values are then added to the cell state ( $\hat{C}_t$ ) by using a tanh layer (green). Eqs. (2) and (3) present these procedures mathematically.

$$i_t = \sigma\left(x_t U^i + y_{t-1} W^i\right) \quad (2)$$

$$\hat{C}_t = \tanh\left(x_t U^g + y_{t-1} W^g\right) \quad (3)$$

The line at the top of the cell is known as the cell state ( $C_t$ ) and has interactions with all components. Information has the opportunity of being forgotten when the old state ( $C_{t-1}$ ) is multiplied by the result of the first forget gate ( $f_t$ ). The product of the second (red) and third (green) gates are then added which results in new information being provided to the cell state and is represented by Eq. (4).

$$C_t = f_t C_{t-1} + i_t \hat{C}_t \quad (4)$$

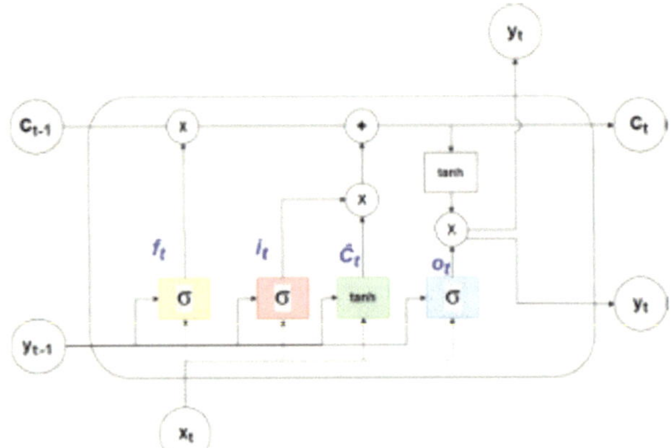


**Figure 2.**
*LSTM network cell.*

Lastly, the output layer of the LSTM cell determines the forecast for the current time step. A sigmoid layer (blue) and tanh layer are multiplied to generate an output ( $y_t$ ). This final step is represented by Eqs. (5) and (6).

$$o_t = \sigma\left(x_t U^0 + y_{t-1} W^0\right) \tag{5}$$

$$y_t = \tanh\left(C_t\right) \times o_t \tag{6}$$

The result of this computational procedure is a time series forecast of future values. However, a large amount of data must be gathered to use as a model input. This data is presented in the next section.

### 3.2 Data required

Historic stream stage height for the location further explained in Section 4 must first be gathered. 113,994 data points were procured that correspond to 15-minute intervals from May 19, 2016 (5 PM) – September 1, 2019 (4 PM). Stage height is herein referred to as 'gauge height' to account for the source of the data. This data is represented graphically in **Figure 3** [22].

Using USGS' flood inundation mapper (FIM), these gauge heights can be tied to a specific flood inundation profile [23]. The FIM is a publicly available tool that provides resulting flood inundation maps for one-foot gauge height increments in image format (.tif). A sliding bar that accomplishes this is available on the online user interface and is presented in **Figure 4**.

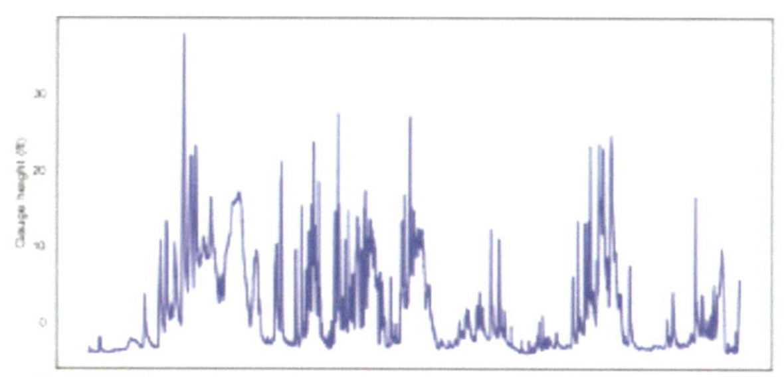

**Figure 3.**
*Stream stage height for example locations.*

**Figure 4.**
*FIM sliding gauge height tool.*

An example of a flash flood inundation profile being uploaded to a GIS software is provided in **Figure 5**. Purple lines correspond to road network data derived from the National Transportation Dataset [24]. Blue raster (grids of pixels) imagery denote the depth of water at discrete locations where darker blue reflects deeper water. Useful geoprocessing techniques that generate actionable decision support tools are presented in the next section.

### 3.3 Geoprocessing procedures

Traffic decisions makers are tasked with identifying flood affected road segments. In **Figure 5**, it can be observed that the flood inundation profile does overlap certain road segments. Relying on visual inspection alone is time consuming and prone to inaccuracies due to human error. A solution to this issue is the application of a set of straightforward geoprocessing tools that are built-in to most GIS softwares: conversion and intersection.

Some tools do not allow raster and vector data layer interoperability. Therefore, it is necessary to convert one of the data layers to establish a consistent data type. One approach is to convert the raster layer into a vector layer using the conversion tool within ArcGIS. **Figure 6** illustrates the result of this operation. The flood inundation profile has been converted into several points at 1-m increments. This spatial resolution can be modified by the user. The road network has been changed from its previous color to improve readability.

Once the raster layer has been converted into vector format, it is eligible for use as an input layer for the intersection tool. The intersection tool generates a point at

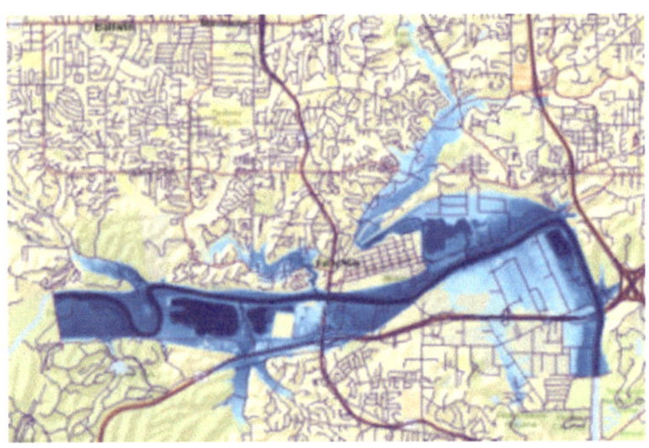

**Figure 5.**
*Flood inundation profile example.*

**Figure 6.**
*Raster layer conversion example.*

every location where there is an intersection between the input layers. In the next section, an illustrative example is provided to demonstrate the effectiveness of the methodology presented.

## 4. Illustrative example

Valley Park, Missouri is located at the intersection of I-44 and State Route 141. This location is the setting for the example figures presented previously. The Meramec River winds through this area and has regularly flooded in recent years. In 2017, the river exceeded its banks and caused significant damage to the surrounding area as seen in **Figure 7**. This location provides a suitable candidate to test the methodology presented given the extent of the flood event and data availability.

First, data is gathered from a nearby stream gauge. **Figure 8** provides a geographical point of reference for the gauge denoted by a green square with respect to I-44 and State Route 141. The data presented in **Figure 5** is then procured and used as an input for the LSTM network. **Figure 9** presents the prediction results of the LSTM model superimposed on the actual data for May 19, 2016-September 1, 2019.

The actual data (blue) can be observed deviating from the prediction results for the training (orange) and testing (green) results of the LSTM network. A lack of discrepancy between the actual data and predictions demonstrates the model's effectiveness. Further, it is useful to determine how the prediction compares with publicly available forecasts for the same location. USGS provides a forecast every six hours. Alternatively, the LSTM network provides 24 predictions in the same period. **Figure 10** provides a comparison of the prediction provided by USGS and the LSTM model for September 1, 2019 (6 PM) – September 3, 2019 (6 AM).

The red line represents the original data. Gauge height is initially observed at just above six feet. From there, it trends in a downwardly direction until it reaches

**Figure 7.**
*Meramec River flood in 2017 [25].*

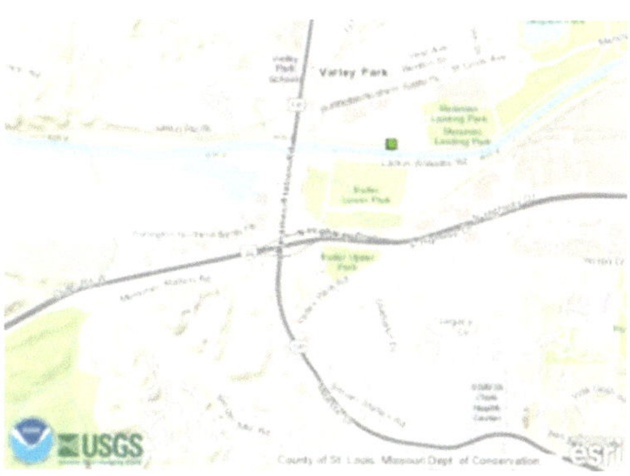


**Figure 8.**
*Gauge location [9].*

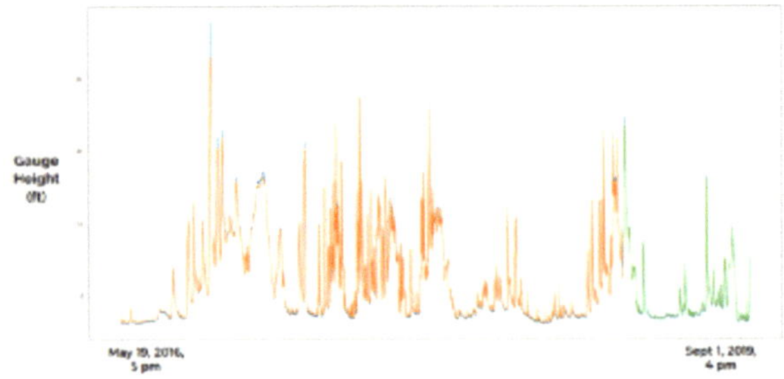


**Figure 9.**
*LSTM training and testing results.*

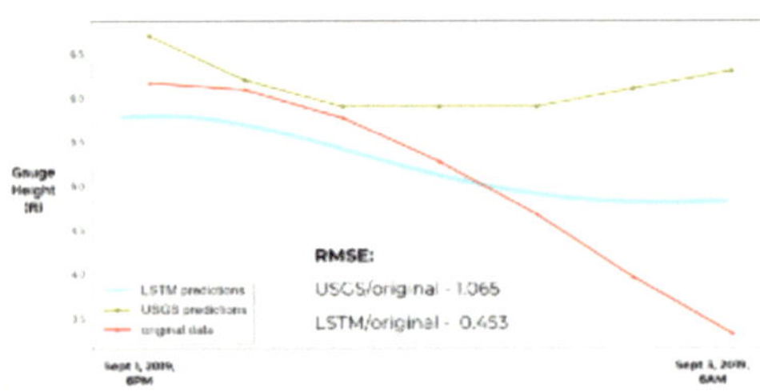


**Figure 10.**
*USGS and LSTM prediction comparison.*

the end of the dataset at less than 3.5 feet. The green line corresponds to the USGS prediction. This prediction initially overshoots the original data before briefly correcting and then diverging significantly from the observed trend. Lastly, the blue line represents the LSTM prediction. At first, this prediction captures the downward trend missed by the USGS prediction. Ultimately, the prediction flattens out and diverges from the original observations but to a lesser extent when compared to the USGS prediction. Root Mean Squared Error (RMSE) values for each of the predictions are provided to further demonstrate the difference in model performance. The RMSE value of 0.453 reported by the LSTM model represents superior accuracy compared to the 1.065 value reported by the USGS prediction. Therefore, the LSTM model presented here improves on the accuracy of publicly available forecasts and can be used as an input for the flood inundation tool.

Valley Park has 43 flood inundation profiles available in one-foot increments from 11–54 feet. The highest stage value recorded at this location is 44.11 feet on December 31, 2015. **Figure 11** provides the flood inundation profile for 45 feet to approximate this event. Note that 45 feet is used instead of 44. This is due to the

**Figure 11.**
*Flood inundation profile for 45 ft. stage value for Valley Park, Missouri.*

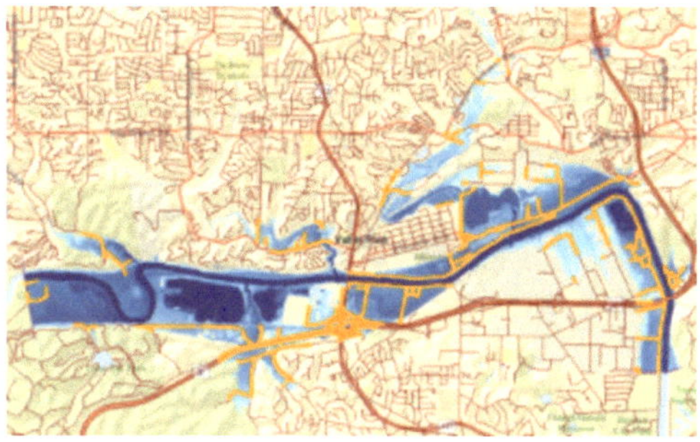

**Figure 12.**
*Flood affected road segments for flood inundation profile corresponding to 45 ft. stage value for Valley Park, Missouri.*

flood inundation profile incremental limitation and opting for a rounding approach that provides a more conservative risk assessment. The inundation profile is then converted to point format and intersected with the road network as illustrated by **Figure 12**.

## 5. Discussion

At present, urban planners such as traffic decision makers rely on static flood inundation maps and post hoc planning to reroute traffic in the event that a flood occurs. This approach puts motorists already in-transit at risk to rapidly changing road conditions. To address these risks, a field of research has emerged to provide decision makers with real-time decision-making tools. However, using time series prediction models that capture river characteristics and integrating them with flood inundation profiles has received limited attention. The methodology provided here addresses this gap.

Traffic decision makers can use the data visualization presented in **Figure 12** as a powerful decision support tool. The flood affected road segments can be easily identified (orange) and rerouting measures can be promptly dispatched. With the improved temporal resolution and accuracy of the LSTM prediction of stage height, traffic decision makers can deploy resources proactively to avoid unnecessary risk to motorists and improve traffic flow. Concluding remarks, limitations, and future work are presented in the next section.

## 6. Conclusion

Flash floods are a frequent and devastating natural disaster. The impetus to manage these events belongs to local decision makers that work in a resource constrained environment. To improve their decision-making effectiveness, a framework was presented that integrates machine learning and geospatial data to extract spatial and temporal trends using publicly available data. An illustrative example was provided to demonstrate the effectiveness of the framework provided. Valley Park, Missouri is located near the intersection I-44 and State Route 141. These roads represent major traffic throughputs and persistent flooding of the Meramec River has jeopardized the safety of motorists and the flow of commercial goods. Using 113, 994 river stage observations procured from a nearby sensor, an LSTM network was developed to improve the accuracy of publicly available forecasts. The result was an improvement in both the frequency and accuracy of forecasts provided.

Once the stage value is predicted it can be tied to a spatial flood inundation profile using the publicly available FIM. Using the flood inundation profile for 45 feet observed at Valley Park as a proxy for the historic crest at this location, data visualization of flood affected road segments was generated in a GIS setting. The key benefit of this output is the ease with which traffic decision makers can use the results presented to inform urban planning and decision making. Traffic decision makers can use the resulting data visualization presented here to guide real-time decision making in the event that a river stage value is predicted to reach a flood event stage for a specified river segment. Despite the usefulness of the findings, there remain a number of model limitations that represent areas of future work.

Model limitations can be divided into two categories: data gathering and model extension. Deep learning models are dependent on large amounts of data. Therefore, sensors that collect data need to be installed and active for an extended period. The cost to install and maintain an enlarged sensor network might be prohibitive for some locations. Due to this fact, model implementation is limited to river locations where sensors are already installed. Additionally, FIM coverage is confined to a small number of locations nationwide. Similarly, to sensor coverage, if there are not already-available flood inundation maps, then the model cannot be applied to those locations. Model extension includes options to improve the model in a material way. One recommendation would be to determine the best locations for road signage that will provide optimal re-routing to motorists given a finite amount of signage. Another approach would involve working with local decision makers to determine re-routing effectiveness based on how quickly resources are deployed given model predictions. Areas of future work not related to model extensions include alternative prediction approaches in river networks with no sensors and refinement of the model to account for flash floods. Each of these components represent considerable opportunity for model enrichment that further improve the decision-making effectiveness for traffic management professionals.

The results presented here demonstrate the utility of using machine learning models and geospatial data to generate data visualization tools that key stakeholders can use to improve planning effectiveness. As data becomes increasingly available, use of comparably sophisticated methods can be applied to a suite of natural disaster phenomena. The outcome of such an undertaking will be the widespread use of data visualization tools that will reduce the risk motorists are exposed to and mitigate the accompanying economic fallout.

## Acknowledgements

This work was partially funded by the Missouri Department of Transportation, Award Number TR201912 and the Mid-America Transportation Center, Award Number 25-1121-0005-130.

## Conflict of interest

The authors declare no conflict of interest.

## Author details

Jacob Hale[1], Suzanna Long[1]*, Vinayaka Gude[2] and Steven Corns[1]

1 Department of Engineering Management and Systems Engineering, Missouri University of Science and Technology, Rolla, MO, United States

2 Department of Arts and Media, Louisiana State University Shreveport, Shreveport, LA, United States

*Address all correspondence to: longsuz@mst.edu

## References

[1] United States Geological Survey. Floods: Things to Know [internet]. 2019. Available from: https://www.usgs.gov/special-topic/water-science-school/science/floods-things-know?qt-science_center_objects=0#qt-science_center_objects [accessed 2019-03-15]

[2] National Severe Storms Laboratory. Flood Basics [internet]. 2019.Available from: https://www.nssl.noaa.gov/education/svrwx101/floods/ [accessed 2019-03-16]

[3] World Health Organization. Floods [internet]Available from: https://www.who.int/health-topics/floods#tab=tab_1 [accessed 2019-05-20]

[4] National Oceanic and Atmospheric Administration. 2010-2019: A landmarkdecade of U.S. billion-dollar weather and climate disasters [internet]. 2020.Available from: https://www.climate.gov/news-features/blogs/beyond-data/2010-2019-landmark-decade-us-billion-dollar-weather-and-climate [accessed 2020-10-15]

[5] National Weather Service. NWS Preliminary US Flood Fatality Statistics[internet]. 2020. Available from: https://www.weather.gov/arx/usflood [accessed 2020-10-15]

[6] Dietsch, B.J., and Strauch, K.R., 2019, Flood-inundation maps of the Meramec River from Eureka to Arnold, Missouri, 2018: U.S. Geological Survey Scientific Investigations Report 2019-5004, 12 p. Available from: https://doi.org/10.3133/sir20195004.

[7] Jha MK, Afreen S. Flooding urban landscapes: Analysis using combined hydrodynamic and hydrologic modeling approaches. Water (Switzerland). 2020;12(7).

[8] Shirowzhan S, Sepasgozar SME, Li H, Trinder J, Tang P. Comparative analysis of machine learning and point-based algorithms for detecting 3D changes in buildings over time using bi-temporal lidar data. Automation in Construction. 2019Sep;105. Available from: https://doi.org/10.1016/j.autcon.2019.102841

[9] Advanced Hydrological Prediction Service. Meramec River at Valley Park [internet]. 2019.Available from: https://water.weather.gov/ahps2/hydrograph.php?wfo=lsx&gage=vllm7[accessed 2019-06-02]

[10] Lago J, De Ridder F, De Schutter B. Forecasting spot electricity prices: Deep learning approaches and empirical comparison of traditional algorithms. Appl Energy [Internet]. 2018;221(February):386-405. Available from: https://doi.org/10.1016/j.apenergy.2018.02.069

[11] Papacharalampous G, Tyralis H. Hydrological time series forecasting using simple combinations: Big data testing and investigations on one-year ahead river flow predictability. J Hydrol [Internet]. 2020;590(May):125205. Available from: https://doi.org/10.1016/j.jhydrol.2020.125205

[12] Liu J, Wang J, Pan S, Tang K, Li C, Han D. A real-time flood forecasting system with dual updating of the NWP rainfall and the river flow. Nat Hazards. 2015;77(2):1161-82.

[13] Adnan, Muhammad, X, Yuan, O, Kisi et al. Application of soft computing models in streamflow forecasting. Proceedings of the Institution of Civil Engineers – Water Management. 2019;172(3):123-124. Available from: https://doi.org/10.1680/jwama.16.00075

[14] Mosavi A, Ozturk P, Chau KW. Flood prediction using machine learning models: Literature review. Water (Switzerland). 2018;10(11):1-40.

[15] Gude V, Corns S, Long S. Flood Prediction and Uncertainty Estimation Using Deep Learning. Water (Switzerland). 2020;12(3).

[16] Bzdok D, Altman N, Krzywinski M. Points of Significance: Statistics versus machine learning. Nat Methods [Internet]. 2018;15(4):233-4. Available from: http://dx.doi.org/10.1038/nmeth.4642

[17] Tehrany MS, Pradhan B, Jebur MN. Flood susceptibility mapping using a novel ensemble weights-of-evidence and support vector machine models in GIS. J Hydrol [Internet]. 2014;512:332-43. Available from: http://dx.doi.org/10.1016/j.jhydrol.2014.03.008

[18] Shafizadeh-Moghadam H, Valavi R, Shahabi H, Chapi K, Shirzadi A. Novel forecasting approaches using combination of machine learning and statistical models for flood susceptibility mapping. J Environ Manage. 2018;217:1-11.

[19] Tien Bui D, Hoang ND, Martínez-Álvarez F, Ngo PTT, Hoa PV, Pham TD, et al. A novel deep learning neural network approach for predicting flash flood susceptibility: A case study at a high frequency tropical storm area. Sci Total Environ [Internet]. 2020;701:134413. Available from: https://doi.org/10.1016/j.scitotenv.2019.134413

[20] Papaioannou G, Vasiliades L, Loukas A. Multi-Criteria Analysis Framework for Potential Flood Prone Areas Mapping. Water Resour Manag. 2015;29(2):399-418.

[21] Publication-ready NN-architecture schematics [internet]. 2021. Available from: alexlenail.me/NN-SVG/index.html [accessed 2021-1-27]

[22] United States Geological Survey. USGS 07019130 Meramec River at Valley Park, MO [internet]. 2019.Available from: https://waterdata.usgs.gov/nwis/uv?site_no=07019130[accessed 2019-10-16] usgs 2019

[23] United States Geological Survey. Flood Inundation Mapper [internet]. 2019.Available from: https://fim.wim.usgs.gov/fim/[accessed 2019-10-18]

[24] United States Geological Survey. USGS National Transportation Dataset (NTD) Downloadable Data Collection [internet]. 2019. Available from: https://catalog.data.gov/dataset/usgs-national-transportation-dataset-ntd-downloadable-data-collectionde7d2[accessed 2019-10-22]

[25] KMOV4. Photos: Before & After Meramec River flooding [internet]. 2017. Available from: https://www.kmov.com/news/photos-before-after-meramec-river-flooding/article_fc16115e-12e2-54e6-bf84-ba1732a4dcbd.html [accessed 2019-10-25]

Chapter 6

# Visual Data Science

*Johanna Schmidt*


## Abstract

Organizations are collecting an increasing amount of data every day. To make use of this rich source of information, more and more employees have to deal with data analysis and data science. Exploring data, understanding its structure, and finding new insights, can be greatly supported by data visualization. Therefore, the increasing interest in data science and data analytics also leads to a growing interest in data visualization and exploratory data analysis. We will outline how existing data visualization techniques are already successfully employed in different data science workflow stages. In some cases, visualization is beneficial, while still future research will be needed for other categories. The vast amount of libraries and applications available for data visualization has fostered its usage in data science. We will highlight the differences among the libraries and applications currently available. Unfortunately, there is still a clear gap between visualization research developments over the past decades and the features provided by commonly used tools and data science applications. Although basic charting options are commonly available, more advanced visualization techniques have hardly been integrated as new features yet.


**Keywords:** visual data science, data visualization, visual analysis, data visualization libraries, data visualization systems

## 1. Introduction

Within the last years, data science has been established as its own important emergent scientific field. Data science is defined as a "concept to unify statistics, data analysis, machine learning, and their related methods" to "understand and analyze actual phenomena with data" [1]. As such, data science comprises more than pure statistical data analytics, but the interdisciplinary integration of techniques from mathematics, statistics, computer science, and information science [2]. Data science also involves the consideration of domain knowledge for the analysis and the interpretation of the data and the results [3].

Data visualization research is largely driven by current use cases that users have to face when working with data. The problems and tasks that need to be solved by data scientists are, naturally, a precious source for further developments in data visualization research. On the other hand, data scientists already use data visualizations to visualize data on a daily basis. It is, therefore, worthwhile to think about how the well-established methods for visual analysis fit into the existing workflows of data scientists [4]. According to recent findings from interviews with people working with data [5], data scientists' tasks usually follow a very similar workflow path, and along this path, different stages can be identified. Every stage poses different challenges for data handling. For example, at the beginning of the workflow, data wrangling is considered to be an essential and tedious part of the workflow. Data wrangling comprises, among others, data parsing, cleaning, and merging. Data visualization techniques can help to quickly identify data flaws like missing data, anomalies like duplicates or outliers, and other inconsistencies in this stage. As a next step, data scientists have to understand the data at hand and evaluate its usefulness for modeling. Here, data visualization can help understand the structure of the data, detect correlations and clusters, and select data parts suitable for modeling.

The rise in data science currently very strongly fuels data visualization techniques by users from very diverse domains. This has now led to many new data visualization tools and libraries being developed. Many of these libraries are open-source and are embedded into programming environments like Python, R, and JavaScript. Prominent examples for such libraries are, for example, *Matplotlib* (Python), *ggplot2* (R), and *D3* (JavaScript). Open source technologies are a great advantage since data scientists can rely on a large community that can provide them with advice and support, and access to a wide range of libraries and plugins. Especially for Python, there are libraries for high-performance computing, numerical calculations, regression modeling, and visualization, which are regularly extended and maintained. On the other hand, feature-rich, standalone visual analysis applications have been increasingly established within the last years. These applications, such as *Tableau*, *Microsoft Power BI* and *Qlik*, provide easy access to data visualization and visual data exploration for users unfamiliar with programming scripting, data wrangling, and/or data visualization design. Standalone applications are usually commercial, since a lot of maintenance and continous development has to happen in the background. As many of these applications are available, data visualization and visual analysis are more widely known and used today in many different domains and are used and applied by many users and domain experts.

This chapter aims to provide a concise overview of existing data visualization techniques for data science and how they fit into the different stages of the data science workflow. Several studies that focused on categorizing and evaluating the different libraries and applications for data visualization currently used in data science will be outlined to create a better picture of which libraries should be used for which type of tasks. Unfortunately, there is still a gap between current research in data visualization and the features and techniques actually provided by libraries and applications. We would, therefore, like to foster the usage of data visualization in data science to bring both communities closer together.

## 2. Visualization supporting data science

Data science is an interdisciplinary approach that combines input from other domains like mathematics, statistics, computer science, or graphics. Given this vast amount of tasks and skills, several studies have been conducted to understand better and start to categorize the tasks and requirements of data scientists. Kim et al. [6] highlighted the diversity of skills, tasks, and toolsets used by data scientists in software development teams. As an important conclusion, they highlighted that the heterogeneity and diversity make it hard to reuse work. Kandel et al. [5] conducted an interview study with several data scientists and categorized them into the three archetypes of *Hacker*, *Scripter*, and *Application User*. Based on the archetype, data scientists use very diverse tools to solve their tasks. The survey by Harris et al. [7] among different data workers, as they call people working with data, from different disciplines, provided a very comprehensive overview of the different tasks data scientists need to solve. As a result of their study, they were then able to categorize data scientists into one of four major categories based on their skills (e.g., business orientation vs. programming skills). In general, both studies concluded that data scientists either prefer to use hands-on scripts and program their own algorithms over using fully-featured applications.

As a basis for better explaining how data visualization fits into the data science workflow, we would like to use the categorization introduced by Kandel et al. [5]. They proposed to divide the data science workflow into five Stages:

- **Discover**: As a first step, data scientists usually search for suitable datasets, either by locating them in databases or online or by asking colleagues. Especially within large organizations, finding and understanding relevant data is often considered a significant bottleneck in the work process, also due to access restrictions.

- **Wrangle**: When available, the datasets need to be brought into the desired format. Data wrangling involves parsing files, manipulating data layouts, and also integrating multiple heterogeneous data sources. Being considered to be a very tedious and highly manual task, data wrangling eats up a majority of the time spent on data analysis.

- **Profile**: After being available in the desired format, the quality of the data has to be verified, and the suitability for the analysis has to be estimated. Datasets often contain severe flaws, including missing data, outlier, erroneous values, and other problems. Understanding the structure of the data is therefore considered an important task in data science.

- **Model**: Finally, an essential and interesting part of the data science workflow is to use the datasets as training sets to train prediction models. In this stage, the models have to be created and evaluated against existing real-world data to test their performance.

- **Report**: All analysis results usually need to be reported to external people, colleagues, or customers. In such a presentation, it is important to cover the essential findings discovered during the data science process. In many cases dashboards or reports are used to present the findings.

In the *Discover* stage, data scientists need to identify the data relevant to their current project. This involves searching for internal but also external data sources. The main challenges in this stage are restricted data access and missing documentation of data attributes. This stage is, in general, not supported by data visualization applications. There are approaches in data visualization to, for example, improve the visualization of search engine results [8], but the general problem of data being difficult to find/access is not treated. We therefore do not concentrate on this stage here in this chapter.

### 2.1 Data wrangling

Data wrangling in the *Wrangle* stage requires to, on the one hand, focus on data flaws like duplicates and inconsistencies (e.g., in naming), and, on the other hand, the process of profiling and transforming datasets. Data wrangling's central goal is to make the data usable in the subsequent steps.

Initially, data wrangling was not considered by data visualization itself, which started to operate once the data was available in the desired format. Since data wrangling nowadays became an essential and tedious task in data science, which eats up a lot of the time in the whole workflow (up to 50–80% [9]), data visualization researchers started to think about techniques how to support this task. *Wrangler* [10] is the most prominent application to mention here, an interactive system for creating data transformations. Changes in the data are visualized, and data scientists can explore the space of possible operations. The *Wrangler* system infers further actions from what has been done by the user so far manually, and in this way, greatly speeds up the wrangling process. The idea was picked up by the company Trifacta, which included the Wrangler idea into their product to build data pipelines.

In general, data wrangling itself constitutes a very interesting use case which, hopefully, in the future, will get more attention by data visualization research. At the moment, data scientists mostly have to rely on manual tasks and scripting tools to get the data into the right format.

## 2.2 Data profiling

The most demanding stage in terms of visualization design is the *Profile* stage, where data scientists need to explore the data to understand its structure. This process is very circular and undirected, without a specific goal in mind. Basic information about the related problem domain is required. The goal is to understand the patterns found in the data. This includes, but is not limited to, the distribution of values, correlations, outliers, and clusters.

Datasets usually contain several quality issues, such as missing values, outliers or extreme values, and inconsistencies. Missing data might be due to observations completely missing in a dataset, which can be in many cases identified by empty cells of *null* values. There might also be cases where numbers encode missing data (e.g., 0 or −1) or characters (e.g., "N/A"), something that needs to be considered during the analysis. Inconsistencies and heterogeneous information are often erroneously created by humans, especially in names and terms, and because certain cells have been overloaded with information. Data scientists need to be aware of these flaws when working with a particular dataset. Checking the quality of a dataset has already been addressed by several approaches in visualization. *Profiler* [11] was intended to support the quality assessment of datasets visually; *Visplause* [12] provided the same for time series data. More generally, Bertini et al. [13] developed quality metrics for multi-dimensional datasets. Quality checks are nowadays also provided as features in standalone visualization applications. In Python, the package *pydqc* provides automatic quality checks.

In data visualization, the process of looking at data from different directions and studying different aspects to understand the data structure [14] is called *exploratory data analysis (EDA)*. EDA requires a high degree of interactivity and interconnectivity between different visualizations from a data visualization perspective. EDA has been studied quite extensively within the last years in visualization research. Several paradigms about interaction design [15] and system design [16] have been established. Exploration of data usually happens by using different views and different visualizations. EDA contrasts with the more traditional statistical approach to data analysis that starts with hypothesis testing. In EDA, data scientists usually do not have a clear goal and should support the hypothesis-building process. The typical EDA tasks [17] are:

- Plotting the raw data,
- Plotting simple statistics (e.g., mean, standard deviation, box plots), and
- Positioning such plots for comparison (e.g., in a multiple-view setting).

These tasks are, in general, supported by all visualization applications. In case scripting languages are used, data scientists tend to create several data representations to check various aspects. Programming environments like *Jupyter* Notebooks [19] enable scientists to combine both data analysis scripts and visualization. An example for showing Python *Plotly* visualizations in *Jupyter* can be seen in **Figure 1**. Such narrative or literate programming tools [20] as notebooks help data scientists to record their steps and decisions in a data analysis workflow. They allow scientists to save whole workflows and, in this way, make decisions and results reproducible. This has also been recognized by data visualization research [21], where researchers increasingly think about new solutions for more advanced data visualization in notebook environments and literate programming.

### 2.3 Modeling

Data scientists make assumptions to find out which types of transformations they need to use for modeling. This also includes understanding which of the data fields are most relevant to a given analysis task. In the *Model* stage, the data is used as input data for building models of the underlying phenomenon. When models have been built, it is important to evaluate them against suitable real-world data.

Building and evaluating simple models like regression models is already supported by some visualization applications [22]. More advanced techniques, often summarized under the term "explainable AI" [23], try to find new, often visual, ways for humans to explain the decision structure of AI (artificial intelligence)

```
import plotly.io as pio
import plotly.express as px
import plotly.offline as py

df = px.data.iris()
fig = px.scatter(df, x="sepal_width", y="sepal_length", color="species", size="sepal_length")
fig
```

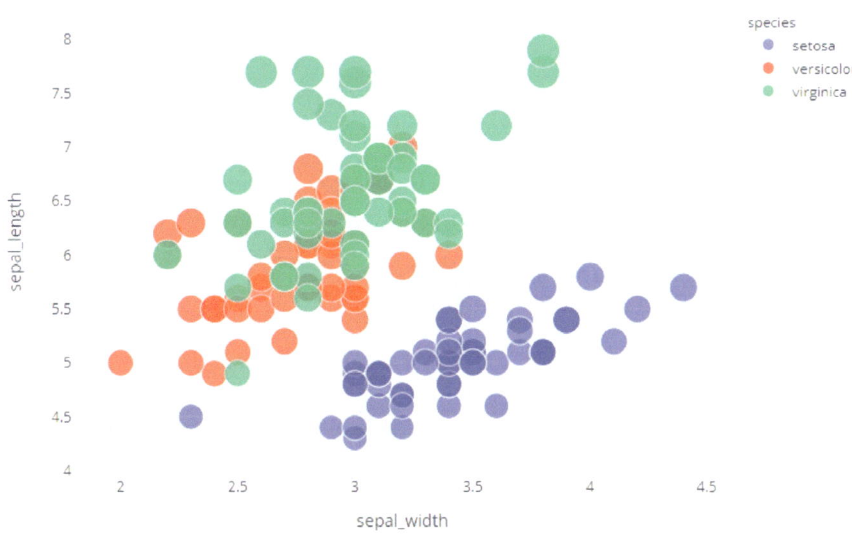


**Figure 1.**
*Jupyter Notebook and Plotly. Literate programming tools as notebooks allow scientists to combine both data analysis scripts and visualization. Figure by [18].*

models and verify their decision according to their own ground-truth knowledge. One problem that is mentioned often by data scientists is the scalability of model testing to large data. Currently, models are often evaluated using EDA techniques very similar to the ones described in the *Profile* stage. In the future, data visualization research will concentrate on advanced methods for the verification of model outputs, especially in relation to the input and output data.

## 2.4 Reporting

In the *Report* stage, mostly simple and easy-to-understand visualizations are needed since here the results of the data analysis stage have to be presented to a broader audience. The use cases in this stage can be mostly covered by employing basic charts, which are already well supported by current data science tools.

In many cases, dashboards with more or less interactivity are used to present the results. Many data science tools already support building dashboards. This was also recognized by the data visualization community recently. Sarikaya et al. [24] pointed out that dashboards are actually much more than just a collection of different graphs and that they need to be treated as separate research objects in data visualization. In their work, they categorized existing dashboards into seven categories, mostly based on the intended task (e.g., information and education vs. decision-making). Three examples are shown in **Figure 2**. Such approaches point

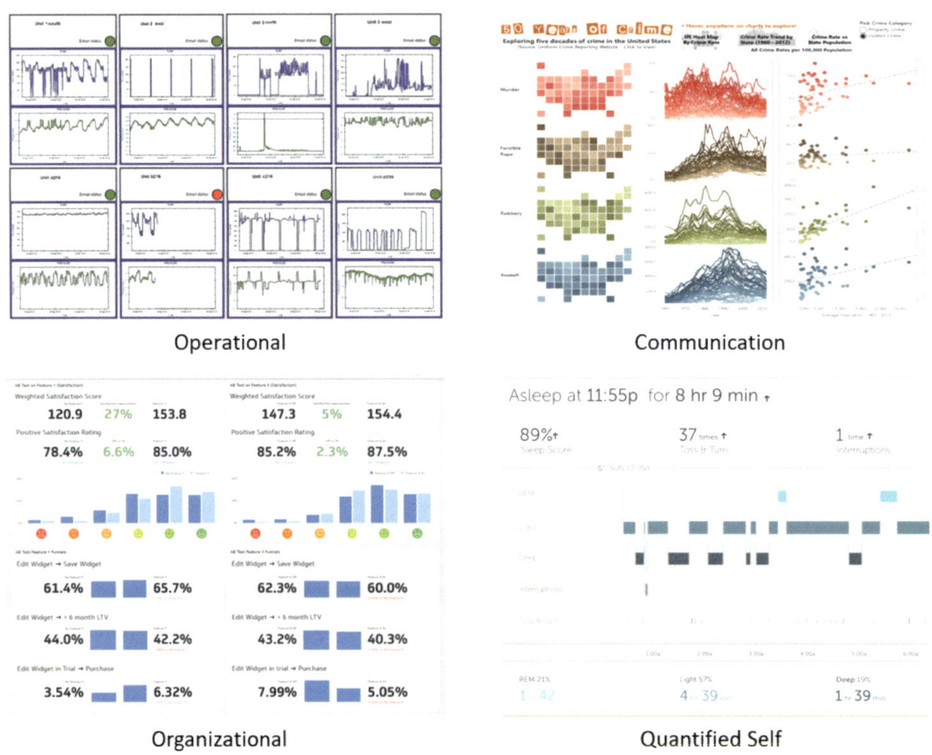


**Figure 2.**
*Dashboards types by [24]. Dashboards for reporting data findings may differ according to the intended user group and task. In this figure, dashboards for operational decision-making, strategic decision-making, communication, and studying your own data (quantified self) are shown. The dashboards in the first column (operational and organizational) target a narrow group of users with particular tasks in mind. The second column's dashboards (communication and quantified self) are intended to be viewed by a larger audience. Images by [24].*

out the necessity for dashboard designers to be clear about the intended user group and always have a clear story when presenting data to external people.

One important aspect to consider is to choose the right visualizations for the right type of data. This is especially important if the exact structures in the data are unknown to the viewers and if the viewers' experience with data visualization is unclear. Based on research on human perception and possibilities for data visualization, researchers started to create guidelines for data- or task-driven suggestions for data visualizations. The *Draco* system by Moritz et al. [25] uses predefined rules to suggest several visualizations based on the data and attempt what should be shown in the data. On their website *From Data to Viz*, Holtz and Healy [26] outline several paths how, starting from a specific data type, certain patterns in the datatype can be visualized. The *Data Visualisation Catalogue* [27] summarizes different visualization techniques and explains how they can be employed to encode information. All-in-all, these approaches show the need for further research on guidelines in data visualization research.

## 3. Data visualization toolboxes

As more and more people started working in data science, more and more software applications for data analysis, many of which are open source, have evolved within the last years [28]. All steps in the data science workflow contain circular processes where data scientists have to rethink actions they made and restart analysis processes from scratch. For this reason of a very interactive and undirected workflow [29], there are no applications, yet, that can cover the entire data science workflow. Data scientists must, therefore, always use a list of combinations of different tools, scripts, and applications to achieve their goals [30]. These tools are often focused on specific tasks, such as efficient data storage and access (e.g., for Big Data applications), data wrangling (i.e., mapping data to another format), or automated analysis (e.g., machine learning). They are based on different programming languages (e.g., Python, R, JavaScript) or are built as fully-featured, standalone applications. In this chapter we specifically concentrate on libraries and tools for data visualization.

When talking about libraries and application for data visualization we use the definition by Rost [31]. They conducted a study about features of visualization libraries and applications by creating one specific chart with different tools. In the study the authors differentiate between *charting libraries* (i.e., programming toolkits) and *apps* (i.e., fully-featured applications). As also noted by Kandel et al. [5], different types of data scientists tend to use different types of tools. A data scientist being identified as an archetype *hacker* would not be happy by having to use a standalone application, because he/she would not be able to access the latest library in a scripting environment, and would therefore not be able to customize his/her individual workflow. We, therefore, stick to this differentiation in this chapter.

### 3.1 Charting libraries

Charting libraries are considered to be all kinds of visualization libraries that need some programming environment to work. In many cases, this is a scripting environment, so many libraries nowadays are based on Python or R. The popularity of data visualization libraries changes from year to year since many of these libraries are open source and therefore undergo continuous adaptations and improvements. Open-source technologies are a great advantage since data scientists can rely on a large community that can provide them with advice and support, and access to a wide range of libraries and plugins. There are some libraries which are repeatedly mentioned in high score lists [32], which are, among others: *ggplot2* (R), *Matplotlib* (Python), *Seaborn* (Python), *Bokeh* (Python), *D3* (JavaScript), *Chart.js* (JavaScript) *Lattice* (R), *Vegas* (Scala), *Breeze-viz* (Scala), *Rgl* (R). The differences between these libraries are, on the one hand, given by the different programming environments they live in. On the other hand, the libraries also offer different features and assets for data visualization. Especially for Python, there are libraries for high-performance computing, numerical calculations, regression modeling, and visualization, which are regularly extended and maintained. This is very similar in the case of R.

The study by Rost [31] reveals fascinating differences between some of the charting libraries. The libraries which have been tested in this study have been classified according to whether they are more suited for analysis tasks or presentation tasks. The results can be seen in **Figure 3**. The analysis very nicely shows that charting libraries for both analysis (*Wrangle*, *Profile*, *Model*) and presentation (*Report*) purposes can be found. Interestingly, the charting libraries rather suited for presentation are based on JavaScript (highlighted by underline). This also shows that web-based visualization methods are currently rather placed in the presentation or reporting phase of a data science workflow. This also makes sense when thinking about the client–server environment of web-based visualizations, and that visualization designers have to carefully think about which type of data to show in this setting–since large datasets could probably not be transferred over the network and could potentially lead to processing or rendering problems on the client-side (e.g., smartphones). Such a careful design can usually only be done after the analysis (*Profile*, *Model*) is already finished.

In the study by Schmidt [33], different charting libraries were compared according to how many different visualization techniques they support. This study revealed big differences between the libraries and identified two leaders in the field, which currently offer the largest range of different visualization techniques. The first leader is *D3* (short for Data-Driven Documents), which is based on JavaScript and uses SVG (Scalable Vector Graphics) elements to display data in the web browser. It was released in 2011 [34] as a successor to the earlier *Protovis* framework to provide a more expressive framework that, at the same time, focuses on web standards and provides improved performance. The second leader is *Plotly*, a collaborative browser-based plotting and analytics platform based on Python [35]. *Plotly* developers especially take care to allow data scientists to share visualizations and information within a large community.

All-in-all, the field of charting libraries is constantly changing, and many more advances are expected to be seen in the future. When deciding for a charting

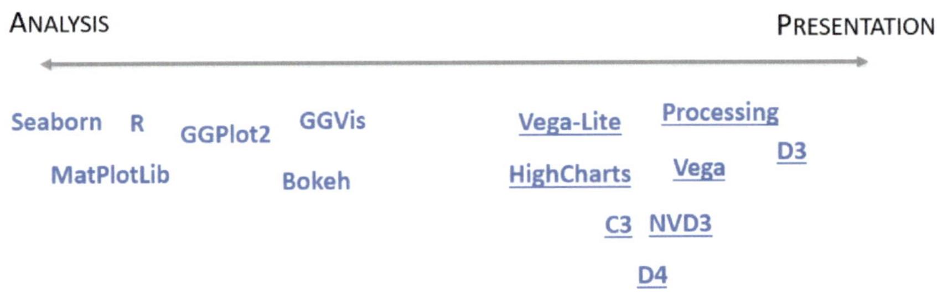


**Figure 3.**
*Comparison of charting libraries. The chart shows the charting libraries used in the study by Rost [31] ranked by whether they are rather suited for analysis or presentation. Charting libraries based on JavaScript (which are, therefore, web-based) are marked by underline. Figure adapted from [31].*

library to be used, other factors like the task to be solved and programming skills have to be considered.

### 3.2 Apps

Apps are considered to be fully-featured, standalone applications. They do not require any programming environment to be installed on a system to run them. Data visualizations can be created by using the user interface tools provided by the application. Apps are more targeted towards users without programming skills who are not familiar with manual data processing, analytics, and visualization. In almost all cases, apps are commercial products. This is because a lot of maintenance and continuous development is needed in the background to keep the apps up-to-date. According to Gartner's Magic Quadrants, a study that is done every year in different areas, the leaders in the field of business intelligence platforms [36] are considered to be *Tableau*, *Microsoft Power BI*, and *Qlik*.

In its yearly study, Gartner compares business intelligence applications that are considered most significant in the marketplace. The applications are evaluated and placed in one of four quadrants, rating the applications as either challengers, leaders, visionaries, or niche players. Many apps are currently available on the market. These applications differ in terms of targeted user groups and also visualization features that they offer. Since Gartner's Magic Quadrants are published every year, interesting patterns can be detected by looking at the yearly changes, as shown in **Figure 4**. The three leaders that have been identified previously already show an excellent performance throughout the last six years. Interestingly, the field of leaders is left to the three main players over the last years. It can also be seen that

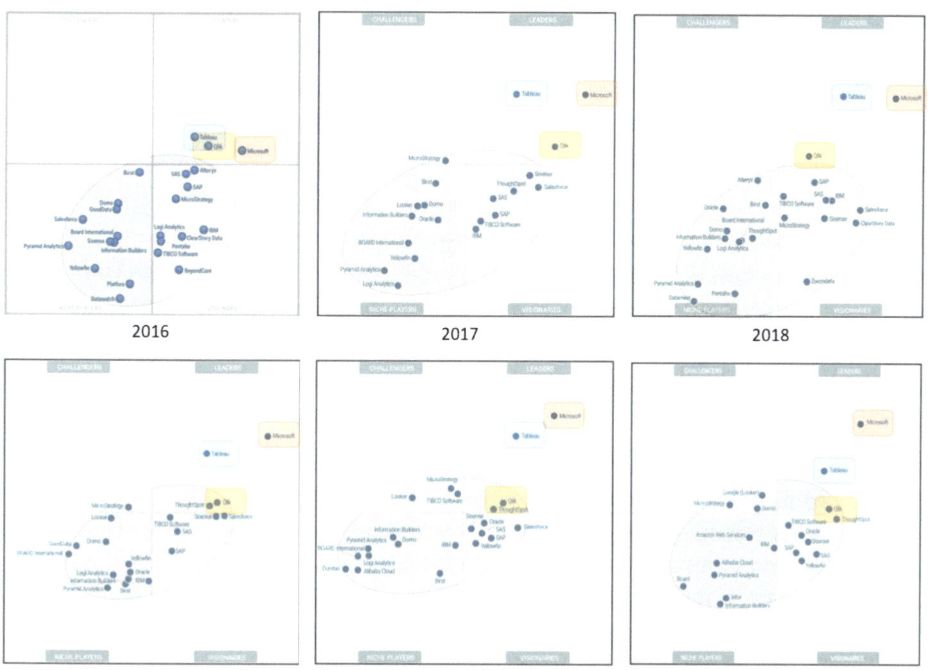


**Figure 4.**
*Gartner's Magic Quadrants of the last six years. The quadrants are divided into the four fields of Leaders, Visionaries, Niche Players, and Challengers. It can be seen that the group of leaders only slightly changed over the last four years. Especially Qlik stayed quite constant. The group of other apps (indicated by a gray circle) shows the field's dynamic movements. New apps have been developed (e.g., Infor, 2021) and others disappeared (e.g., GoodData, 2019). Figure adapted from [37].*

other tools appeared or vanished over the years, which shows the dynamic of the market of business intelligence tools.

The differences between commercial tools have also been highlighted in other studies. Zhang et al. [38] concentrated on specific visualization techniques and evaluated their usage in commercial business analytics tools. They ranked tools and applications based on classifications according to feature richness, flexibility, learning curve, and tasks (e.g., for analysis or presentation). Behrisch et al. [39] conducted an exhaustive survey on commercial visual analytics tools, evaluating them according to which degree they feature data handling, visualization, and automated analysis. Their findings classified the applications according to whether they are more suited for presentation or exploratory analysis. The results show that basically all applications feature data presentation, which is mainly supported by creating dashboards. Some of the applications like *Tableau* or *Qlik* also provide the ability to publish web-based dashboards. Interestingly, only about 50% of the applications were identified to be suited for exploratory analysis (like *Tableau*, *TIBCO Spotfire*, or *Microsoft Power BI*). The authors also identified the applications as useful for different types of users, mainly upper management, reporting managers, or data analysts.

The vast amount of libraries and tools being available has inspired researchers to conduct studies for quantifying, evaluating, and ranking tools and applications that data scientists use. Gartner's Magic Quadrant and several studies about apps for data visualization in data science show no tools that cover all tasks and needs. The selection of an app to be used mainly depends on the tasks that need to be solved (e.g., analysis vs. presentation) and on the scope where the app should be used in.

## 4. Integration of visualization

Visualization researchers were very successful within the last decades, generating many different novel techniques for the visual representation of data. These techniques range from approaches for the efficient representation of data (e.g., parallel coordinates) to proposed interaction and user guidance workflows (e.g., overview-first, details-on-demand). Current surveys show a large variety of visualization techniques. A survey of survey papers in information visualization by McNabb and Laramee [40] classified already over 80 survey papers describing relevant state-of-the-art techniques, and a more recent survey of books in information visualization revealed a similar quantity and variety [41]. Unfortunately, there is only minimal overlap between the recent developments in visualization research and the data visualization features offered by charting libraries and apps. Most of the tools and applications feature basic charts and plots (e.g., scatter plots, bar charts, bubble charts, radar charts), but more advanced visualization techniques (e.g., chord diagrams, horizon graphs) can hardly be found.

This was confirmed by several studies on the integration of visualization techniques in common libraries and applications. Harger and Crossno [42] evaluated the feature richness of open source toolkits for visual analytics. They evaluated the toolkits used for the study based on which basic chart types (e.g., bar charts, line charts), which types of graph visualization (e.g., circular or force-directed layouts), and which types of geo-spatial visualization techniques (e.g., choropleth maps, cartograms) they feature. They concluded that some toolkits are more targeted towards analytics, and some are more targeted towards visualization. Like this study, Schmidt [33] surveyed commonly used tools and applications and evaluated the visualization techniques they feature. They focused on visualization techniques rather than on derived attributes (e.g., feature richness) and included more recent advances in visualization research, considered open source tools and commercial applications to produce a complete picture of visualization techniques usage. They also concentrated on 2D information visualization techniques, as these techniques are more relevant for data science and data analytics, and disregarded spatial techniques like 3D volume rendering.

In all studies that have been conducted so far, not surprisingly, basic chart types like scatter plots and bar charts are highly supported by all evaluated tools and applications. From the more advanced visualization techniques, multi-dimensional techniques like parallel coordinates and radar charts are already widely used and known and therefore included in many of the tools. The same applies to scatter plot matrices and heatmaps. Techniques for hierarchical data are also well supported, especially by open source tools. Visualization techniques for temporal data are not available in the majority of the tools and applications. This is probably because temporal data (e.g., time-series data) is a particular data type used only for specific tasks. Users usually use their own tools for these purposes. Therefore, temporal data techniques have not been included yet in common tools and applications, as these tools usually try to address a broader range of data scientists and data analysts. Some visualization techniques have not been integrated into any tool or application yet, like time nets, data vases, or people garden.

From a tools and applications point of view, *Plotly* and *D3* notably provide the most features among all the tested open source tools. Other tools are targeted towards exceptional functionalities, like *dygraphs* for scientific plots, which only feature a minimal range of visualization techniques. Other libraries which are intended to be used in web-based applications (e.g., *Chart.js* or *Google Charts*) feature only visualization techniques that will most likely be needed in a web-based context. Open source tools, especially *ggplot2*, benefit a lot from the community's input since many advanced visualization techniques are only featured via extensions. In the group of commercial tools, it can be depicted that *Tableau*, *Microsoft Power BI*, and *Highcharts* feature most of the hereby evaluated visualization techniques.

Data scientists could be supported in all stages of their workflow by using visual tools. Interestingly, visualization techniques are currently mostly applied in the *Report* stage, at the end of the data science workflow. This stands in contrast to the fact that interactive data exploration workflows are strongly promoted by visualization research. Even worse, the support for more advanced visualization techniques, especially for interactive data exploration, is still minimal. This has been identified as the "Interactive Visualization Gap" by Batch and Elmquist [43]. Further exchange with data science is considered a valuable and important goal for the visualization community. Previous research efforts in data science revealed that the gap between new developments in visualization research and their application "in the wild" still exists and will hopefully be further mitigated in the future.

## 5. Conclusions

Data visualization can provide substantial support for users working with data. Data visualization techniques have proven to be useful for different steps in the data science workflow. The techniques differ in the interactivity and complexity of the representations. Many of the visualization techniques have been successfully integrated into libraries and applications for data visualization. Especially in the open-source sector, many new directions have opened up within the last years. Due to the programming languages' success, like Python, R, and Scala, libraries targeted towards these programming environments are becoming especially popular.

Among them are *Plotly* for Python and *ggplot2* for R. Also, web-based applications increasingly gain importance. That is why JavaScript-based libraries like *D3* and *Chart.js* can also be found among the most popular data visualization libraries. The market of business intelligence tools is also very dynamic, but it shows some three leaders in the field, namely *Tableau*, *Microsoft Power BI* and *Qlik*. Different types of data scientists require different libraries or applications. It, therefore, can be seen that applications are increasingly targeted towards a specific goal and are designed to solve specific types of tasks. However, when looking at the data visualization techniques offered by the most prominent libraries and applications, the "Interactive Visualization Gap" for exploratory data analysis still exists. Many recent developments and implementations in data visualization research do not find their way into existing libraries and applications. Therefore, the further exchange between data science and data visualization is highly recommended, as both parties can learn a lot from each other and, together, further foster the usage of data visualization in data analytics.

## Acknowledgements

VRVis is funded by BMK, BMDW, Styria, SFG, Tyrol and Vienna Business Agency in the scope of COMET - Competence Centers for Excellent Technologies (879730) which is managed by FFG.

## Author details

Johanna Schmidt
VRVis Zentrum für Virtual Reality und Visualisierung Forschungs-GmbH, Vienna, Austria

*Address all correspondence to: johanna.schmidt@vrvis.at

## References

[1] Chikio Hayashi (1998) *What is Data Science? Fundamental Concepts and a Heuristic Example*. In Data Science, Classification, and Related Methods, pp. 40—51. Springer Japan.

[2] Mark A. Parsons, Øystein Godøy, Ellsworth LeDrew, Taco F. de Bruin, Bruno Danis, Scott Tomlinson, and David Carlson (2011) *A conceptual framework for managing very diverse data for complex, interdisciplinary science*. Journal of Information Science, 37(6): 555—569.

[3] David M. Blei and Padhraic Smyth (2017) *Science and Data Science*. Proceedings of the National Academy of Sciences, 114(33):8689—8692.

[4] Natalia Andrienko, Gennady Andrienko, Georg Fuchs, Aidan Slingsby, Cagatay Turkay, and Stefan Wrobel (2020) *Visual Analytics for Data Scientists*. Springer International Publishing.

[5] Sean Kandel, Andreas Paepcke, Joseph M. Hellerstein, and Jeffrey Heer (2012) *Enterprise Data Analysis and Visualization: An Interview Study*. IEEE Transactions on Visualization and Computer Graphics, 18(12):2917–2926.

[6] Miryung Kim, Thomas Zimmermann, Robert DeLine, and Andrew Bege (2018) *Data Scientists in Software Teams: State of the Art and Challenges*. IEEE Transactions on Software Engineering, 44(11):1024—1038.

[7] Harlan D. Harris, Sean P. Murphy and Marck Vaisman (2013) *Analyzing the Analyzers: An Introspective Survey of Data Scientists and Their Work*. O'Reilly Media, Inc.

[8] Edward Clarkson, Krishna Desai, and James Foley (2009) *ResultMaps: Visualization for Search Interfaces*. IEEE Transactions on Visualization and Computer Graphics 15(6):1057–1064.

[9] Tye Rattenbury, Joseph M. Hellerstein, Jeffrey Heer, Sean Kandel, and Connor Carreras (2017) *Principles of Data Wrangling: Practical Techniques for Data Preparation*. O'Reilly Media, Inc.

[10] Sean Kandel, Andreas Paepcke, Joseph Hellerstein, and Jeffrey Heer (2011) *Wrangler: Interactive visual specification of data transformation scripts*. In Proceedings of the SIGCHI Conference on Human Factors in Computing Systems, CHI '11, May 7–12, Vancouver, Canada, pp. 3363–3372.

[11] Sean Kandel, Ravi Parikh, Andreas Paepcke, Joseph M. Hellerstein, and Jeffrey Heer (2012) *Profiler: Integrated Statistical Analysis and Visualization for Data Quality Assessment*. Proceedings of the International Working Conference on Advanced Visual Interfaces, AVI '12, May 22–25, Capri, Italy, pp. 547—554.

[12] Clemens Arbesser, Florian Spechtenhauser, Thomas Mühlbacher, and Harald Piringer (2016) *Visplause: Visual data quality assessment of many time series using plausibility checks*. IEEE Transactions on Visualization and Computer Graphics 23(1):641–650.

[13] Enrico Bertini, Andrada Tatu, and Daniel Keim (2011) *Quality Metrics in High-Dimensional Data Visualization: An Overview and Systematization*. IEEE Transactions on Visualization and Computer Graphics 17(12):2203–2212.

[14] Peter Filzmoser, Karel Hron, and Matthias Templ (2018) *Exploratory Data Analysis and Visualization*. Applied Compositional Data Analysis: With Worked Examples in R, pp. 69–83, Springer International Publishing.

[15] Arvind Satyanarayan, Kanit Wongsuphasawat, and Jeffrey Heer

(2014) *Declarative Interaction Design for Data Visualization*. Proceedings of the 27th Annual ACM Symposium on User Interface Software and Technology, UIST ‘14, Honolulu, Hawaii, Oct 5–8, USA, pp. 669—678.

[16] Tamara Munzner (2014) *Visualization Analysis and Design*. Taylor & Francis, Inc.

[17] NIST/SEMATECH e-Handbook of Statistical Methods (2021) http://www.itl.nist.gov/div898/handbook/ [Accessed 2021-03-01].

[18] Project Jupyter (2021) *Interactive data visualizations* https://jupyterbook.org/interactive/interactive.html [Accessed 2021-03-02].

[19] Project Jupyter (2021) https://jupyter.org/ [Accessed 2021-03-02].

[20] Mary Beth Kery, Marissa Radensky, Mahima Arya, Bonnie E. John, and Brad A. Myers (2018) *The Story in the Notebook: Exploratory Data Science Using a Literate Programming Tool*. In Proceedings of the CHI Conference on Human Factors in Computing Systems, CHI ‘18, Apr. 21–26, Montreal QC, Canada.

[21] Yifan Wu, Joseph M. Hellerstein, and Arvind Satyanarayan (2020) *B2: Bridging Code and Interactive Visualization in Computational Notebooks*. In Proceedings of the 33rd Annual ACM Symposium on User Interface Software and Technology, UIST ‘20, Oct 20–23, Virtual Event, pp. 152–165.

[22] Thomas Mühlbacher and Harald Piringer (2013) *A partition-based framework for building and validating regression models*. IEEE Transactions on Visualization and Computer Graphics 19 (12):1962–1971.

[23] Wojciech Samek, Grégoire Montavon, Andrea Vedaldi, Lars Kai Hansen, and Klaus-Robert Müller (2019) *Explainable AI: Interpreting, Explaining and Visualizing Deep Learning*. Springer International Publishing.

[24] Alper Sarikaya, Michael Correll, Lyn Bartram, Melanie Tory, and Danyel Fisher (2019) *What Do We Talk About When We Talk About Dashboards?*. IEEE Transactions on Visualization and Computer Graphics 29(1):682—692.

[25] Dominik Moritz, Chenglong Wang, Gregory Nelson, Halden Lin, Adam M. Smith, Bill Howe, Jeffrey Heer (2019) *Formalizing Visualization Design Knowledge as Constraints: Actionable and Extensible Models in Draco*. IEEE Transactions on Visualization and Computer Graphics 25(1):438–448.

[26] Yan Holtz and Conor Healy (2017) Holtz, Y. and Healy, C. (2017) *From Data to Viz* https://www.data-to-viz.com/ [Accessed 2021-02-20].

[27] Severino Ribecca (2021) *The Data Visualisation Catalogue*. https://datavizcatalogue.com/ [Accessed 2021-03-02].

[28] Panagiotis Barlas, Ivor Lanning, and Cathal Heavey (2020) *A survey of open source data science tools*. International Journal of Intelligent Computing and Cybernetics 8(3):232–261.

[29] Jiali Liu, Nadia Boukhelifa, and James R. Eagan (2019) *Understanding the Role of Alternatives in Data Analysis Practices*. IEEE Transactions on Visualization and Computer Graphics 26(1):66—76.

[30] Sara Alspaugh, Nava Zokaei, Andrea Liu, Cindy Jin, and Marti A. Hearst (2019) *Futzing and Moseying: Interviews with Professional Data Analysts on Exploration Practices*. IEEE Transactions on Visualization and Computer Graphics 25(1):22—31.

[31] Lisa Charlotte Rost (2016). *What I Learned Recreating One Chart Using 24*

*Tools*. https://source.opennews.org/articles/what-i-learned-recreating-one-chart-using-24-tools/ [Accessed 2021-03-05].

[32] Bob Hayes (2019) *Business Broadway: Programming Languages Most Used and Recommended by Data Scientists*. https://businessoverbroadway.com/2019/01/13/programming-languages-most-used-and-recommended-by-data-scientists/ [Accessed 2021-02-21].

[33] Johanna Schmidt (2020) *Usage of Visualization Techniques in Data Science Workflows*. In Proceedings of the 15th International Joint Conference on Computer Vision, Imaging and Computer Graphics Theory and Applications, VISIGRAPP '20, Valletta, Malta, Feb 27–29, pp. 309–316.

[34] Michael Bostock, Vadim Ogievetsky, and Jeffrey Heer (2011) *D3: Data-Driven Documents*. IEEE Transactions on Visualization and Computer Graphics 17(12):2301—2309.

[35] Plotly Technologies Inc (2015) *Collaborative data science*. Montréal, QC.

[36] Gartner (2021) *Magic Quadrant for Analytics and Business Intelligence Platforms*. https://www.gartner.com/reviews/market/analytics-business-intelligence-platforms [Accessed 2021-03-01].

[37] Gartner (2021) *Gartner Magic Quadrant*. https://www.gartner.com/en/research/methodologies/magic-quadrants-research [Accessed 2021-02-05]

[38] Leishi Zhang, Andreas Stoffel, Michael Behrisch, Sebastian Mittelstadt, Tobias Schreck, René Pompl, Stefan Hagen Weber, Holger Last, and Daniel Keim (2012) *Visual analytics for the big data era – A comparative review of state-of-the-art commercial systems*. In Proceedings of the IEEE Conference on Visual Analytics Science and Technology, VAST '12, Oct. 14–19, Seattle, WA, USA, pp. 173—182.

[39] Michael Behrisch, Dirk Streeb, Florian Stoffel, Daniel Seebacher, Brian Matejek, Stefan Hagen Weber, Sebastian Mittelstaedt, Hanspeter Pfister, and Daniel Keim (2018) *Commercial Visual Analytics Systems-Advances in the Big Data Analytics Field*. IEEE Transactions on Visualization and Computer Graphics 25(1):3011–3031.

[40] Liam McNabb and Robert S. Laramee (2017) *Survey of Surveys (SoS) - Mapping The Landscape of Survey Papers in Information Visualization*. Computer Graphics Forum 36:589—617.

[41] Dylan Rees and Robert S. Laramee (2019) *A Survey of Information Visualization Books*. Computer Graphics Forum, 38:610—646.

[42] John R. Harger and Patricia J. Crossno (2012) *Comparison of Open Source Visual Analytics Toolkits*. Proceedings of SPIE - The International Society for Optical Engineering, 8294.

[43] Andrea Batch and Niklas Elmqvist (2018) *The Interactive Visualization Gap in Initial Exploratory Data Analysis*. IEEE Transactions on Visualization and Computer Graphics, 24(1):278—287.